Praise for Andreas Wagner

Arrival of the Fittest:

'Fundamental. Entertaining. Brilliant.'
Rolf Dobelli, author of *The Art of Thinking Clearly*

'Andreas Wagner is one of those rare scientists with the courage and intellect to see the real nature of evolution.'
Frank Vertosick, author of *When the Air Hits Your Brain*

'*Arrival of the Fittest* should be mandatory, corrective reading... mind-bending... tremendously exciting.'
BBC Focus

'Quite astounding... the ideas are big, and the numbers hyper-astronomical, but Wagner has a gift for explaining the abstract...elegantly.'
Times Higher Education

'An eye-opener... clear and elegant, with vivid analogies and concrete examples... You'll never think about evolution in the same way again.'
New Scientist

'A truly revolutionary book.'
Independent

Life Finds A Way:

'A wonderful, mind-expanding book. Prepare to be surprised, enlightened and awed.'
Alice Roberts, author of *Ancestors*

'In this remarkably wide-ranging book, Andreas Wagner shows what nature can teach us about creativity, and his answers hold and important message for the way we educate our children and run our institutions.'
Philip Ball, author of *Beyond Weird*

'*Life Finds a Way* weaves a coherent and compelling narrative about how nature achieves creativity. Not only that, we also learn how to cultivate creativity in our own lives.'
George Dyson, author of *Turing's Cathedral*

Also by Andreas Wagner

Life Finds a Way: What Evolution Teaches Us About Creativity
The Arrival of the Fittest: How Nature Innovates
The Origins of Evolutionary Innovations
Paradoxical Life

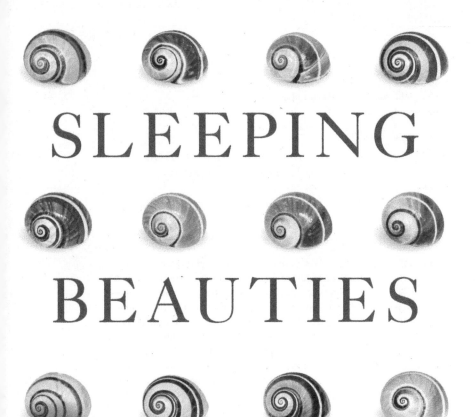

SLEEPING

BEAUTIES

THE MYSTERY OF DORMANT
INNOVATIONS IN NATURE
AND CULTURE

ANDREAS WAGNER

ONEWORLD

A Oneworld Book

First published by Oneworld Publications in 2023
Reprinted, 2023

Illustration credits: Fig. 1 *Welwitschia Mirabilis* © evenfh/Shutterstock;
Fig. 2 *E. coli* © National Institute of Allergies and Infectious Diseases under CCBY20; Fig. 3.1 and 3.2 provided by the author; Fig. 4 *Archaeopteryx* © Mark Brandon/Shutterstock; Fig. 5 *Fustiglyphus annulatus* © James
St John under CCBY20; Fig. 6 *Daphnia Magna* © Hajime Watanabe,
under CCBY20; Fig. 7 *Deinopidae* © Dr Morley Reader/Shutterstock; Fig.
8 James Lind courtesy of Wellcome Collection under CCBYSA40; Fig.
9 Early pacemaker courtesy of Wellcome Collection under CCBYSA40;
Fig. 10 Penicillin poster courtesy of Wellcome Collection under CCBY-
SA40; Fig. 11 Hook-nosed sea snake © Raimonds Romans raymoonds/
Shutterstock

ISBN 978-0-86154-527-8
eISBN 978-0-86154-528-5

Typeset by Geethik Technologies
Printed and bound in Great Britain by Clays Ltd, Elcograf S.p.A.

Oneworld Publications
10 Bloomsbury Street
London WC1B 3SR
England

For my father,
in gratitude

CONTENTS

Introduction: Names writ in water   1

Part One: Nature

Chapter 1   Instant innovation   17

Chapter 2   The long fuse   38

Chapter 3   Molecular motorways   74

Chapter 4   Good vibrations   95

Chapter 5   The birth of genes   123

Part Two: Culture

Chapter 6   The crow and the pitcher   149

Chapter 7   Numbering neurons   165

Chapter 8   Concealed relationships   185

Chapter 9   Reinventing the wheel   197

Chapter 10   Sleeping beauties   221

Acknowledgements   239

Notes   241

Bibliography   281

Index   327

Many a jewel of untold worth
Lies slumbering at the core of earth,
In darkness and oblivion drowned;

Many a flower has bloomed and spent
The secret of its passionate scent
Upon the wilderness profound.

(Charles Baudelaire, *Le Guignon*[1])

INTRODUCTION
NAMES WRIT IN WATER

QUICK, WHAT ARE THE MOST successful organisms on the planet? Many people will answer with apex predators like lions and great white sharks. Others will bring up birds, insects or bacteria. But few people will mention a family of plants that is right up there with the best: grasses. These organisms meet at least two criteria for spectacular success. The first is abundance. Grasses cover the North American prairies, the African savannahs, the Eurasian steppes and countless other grasslands. The Eurasian steppes alone span eight thousand kilometres, from the Caucasus to the Pacific Ocean. A second criterion is the number and diversity of species. Since the time grasses originated in life's evolution, they have evolved into ten thousand species with an astonishing variety of forms, from centimetre-high tufts of hair grass adapted to the freezing cold of Antarctica to the towering grasses of northern India that can hide entire elephant herds, and to Asian bamboo forests, with 'trees' that grow up to thirty metres tall.

But grasses weren't always so spectacularly successful. For tens of millions of years – most of their evolutionary history in fact – grasses barely eked out a living. They failed to flourish by any standard.

The origin of grasses dates back to the age of dinosaurs, more than sixty-five million years ago. But for many million years their fossils are so rare that they cannot possibly have been abundant. And they did not become today's dominant species until less than twenty-five million years ago, more than forty million years *after* their origin.

Why did grasses have to wait forty million years for their pro-verbial spot in the sun? This mystery deepens once you know that evolution endowed grasses with multiple survival-enhancing innovations right from the start. Among them are chemical de-fences like lignin and silicon dioxide that grind down the teeth of grazing animals. They also protect grasses against drought, as do sophisticated metabolic innovations that help grasses con-serve water.

With these and other innovations you'd think that grasses would do very well. But they didn't, for the unimaginably long time of forty million years. And their delayed success holds a profound truth about new life forms. Success depends on much more than some intrinsic characteristic of a new life form, some inner *quality*, like an enhancement or a novel ability bestowed by an innovation. It depends on the world into which this life form is born.

Grasses are not unique in this way. They are among myriad new life forms whose success – measured in abundance or di-versity of species – was delayed for millions of years. The first ants, for example, appear on the scene 140 million years ago. However, ants did not begin to branch into today's more than eleven thousand species until forty million years later. Mammals with various lifestyles – ground-dwelling, tree-climbing, flying or swimming – originated more than a hundred million years before they became successful sixty-five million years ago. And a family of salt water clams had to wait for a whopping 350 million years before it hit the big time, diversifying into 500 species.

These and many other new life forms remained dormant before succeeding explosively. They are the sleeping beauties of biological evolution. They fascinate me to no end, because they cast doubt on the truths about success and failure that we hold self-evident. And these doubts apply not just to the innovations of nature but also to those of human culture.

When life first crawled out of the primordial soup, when it first discovered how to extract energy from minerals, from organic molecules and from sunlight, when it first learned to swim for a living through vast primordial oceans, when it first formed multicellular organisms in which highly specialised cells share the labour and sacrifice of growing and reproducing, of escaping predators and stalking prey, of self-defence and attack, when it mastered each of these challenges, it had to innovate. And each challenge can be met in many ways, each emerging as a creative product of biological evolution, each embodied in a species with a unique lifestyle, millions of them and counting as evolution marches on.

Innovation did not stop with biological evolution. Species with sophisticated nervous systems like chimpanzees, dolphins and crows have discovered simple technologies, tools they use to hunt or gather food. In the ten thousand years since the agricultural revolution, human culture has come up with revolutionary innovations such as mathematics and writing, as well as countless smaller ones, from the wheel to wallpaper. Human ingenuity has discovered fundamental laws of nature and produced myriad creative works, from poems to songs, symphonies and novels. Countless sleeping beauties are among them. They include ignored breakthrough technologies like radar, neglected scientific discoveries like Gregor Mendel's genetic laws of inheritance, or artistic works like Vermeer's painting *Girl with a Pearl Earring*, which languished unrecognised for more than a century.

Granted, nature and culture do not create in exactly the same way. The ink and paper of Newton's *Principia* is a different substrate of creativity than the cells, tissues and organs in a blue whale. A writer's grit in wrestling with the fifteenth draft of a chapter is a different motor of creation than random mutations of DNA. A patent's commercial value is a different measure of success than how often *Escherichia coli* divides every day. But beyond these differences lie deeper similarities. One of them is that a great number of innovations arrive before their time. Sleeping beauties, creative products without apparent merit, value or utility, but with the power to transform life given enough time, are everywhere in both nature and culture. They will help us understand that Gregor Mendel's ignored laws of inheritance and Johannes Vermeer's forgotten paintings are part of a broad pattern in a history of innovation that goes back all the way to the origin of life. The sleeping beauties of nature can help us understand why creating may be easy, but creating *successfully* is beyond hard. It is outside the creator's control.

I am a biologist and my life's aspiration is to understand how biological evolution creates new solutions to life's problems. In this goal I am supported by a team of like-minded young researchers in my laboratory at the University of Zürich. Some of these researchers evolve organisms in the laboratory, where evolution innovates right before our eyes. Others analyse reams of DNA data to understand the origins of unique lifestyles in many organisms. Yet other researchers use the abstract language of mathematics to search for universal laws behind evolution's creative power. All of us want to understand how nature creates. And we are blessed to live in a time where this distant goal is closer than ever. We benefit from a revolution of molecular biology that was launched in the 1950s with the discovery of DNA's double-helical staircase, and that continues unabated in the twenty-first century. It reveals more of life's secrets day after day. These secrets revolve around the trillions of molecules that co-operate in our cells, tissues and organs to keep us alive, molecules that form the foundation of every single innovation during life's evolution, including the origin of life itself.

Whenever a mutation alters an organism's DNA, it ultimately alters some of the molecules that this DNA encodes. It alters what they do, how fast they do it, and how well they do it. These alterations are ultimately responsible for everything new that evolution has created, from the whirling flagella of bacteria to the razor-sharp eyes of falcons and the neural wiring of our brains that make language, arts and science possible. When my research associates study evolution's creations, they study these molecules, how they change over time, and how this change creates new forms of life. And they discover sleeping beauties everywhere. Among them are bacterial enzymes that have evolved for one job but can perform multiple others, such as to cleave and destroy synthetic antibiotics that do not occur in nature. Their skills remained useless until biochemists discovered these antibiotics and doctors used them to fight bacteria. Also among them are entirely new genes that originate spontaneously in large numbers while genomes – including our own – evolve. Each such gene is a solution in search of a problem, which may arise long after the gene's origin, or never.

This book is also about these discoveries. By diving deeply into this molecular realm, we can understand natural innovations far more deeply than through fossils alone. We can understand not just why nature is so innovative, but also how it innovates. And we can understand why so many of its innovations are – must be – dormant.

This dormancy does not shed new light on nature's creative potential alone. Because the innovations of nature are so different from those of human culture, the commonalities between them harbour lessons about all creative processes. These lessons are the subject of this book.

One of these lessons is that innovation – quiescent or not – comes easily to both nature and culture. That is not self-evident. Experts still debate how Darwinian evolution can bring forth the truly new, because natural selection can select only what is already there. On its own, it cannot create new forms of life. Intelligent design creationists even argue that true innovation is impossible in biological evolution. The examples in this book prove the opposite.

They begin by showing that innovation is not precious and rare but frequent and cheap. Chapter 1 highlights underappreciated facts about evolution that illustrate how easy innovation really is. One of them is that most biological innovations are no singularities in life's history. Most were discovered two, three or many times. One example is agriculture, not that of humans but that of ants. Just like human farmers cultivate plants, ants cultivate fungi. They create elaborate fungal gardens, which they groom by removing waste and weeding out unwanted fungal species. They even medicate their fungal crops with antibiotics like streptomycin to combat harmful bacteria.[1]

The sheer speed of evolution's response to environmental change makes the same point. Evolution can respond almost instantly even to transformative change like the rapid rising of the Andes or the emergence of giant Lake Malawi, where dozens or hundreds of new species have evolved and colonised a new environment in a few million years.

Evolution experiments like those we perform in Zürich tell the same story. When we and other researchers expose organisms to a new environment – even a near-lethal one – most evolve to survive and thrive rapidly, within mere weeks. Life usually does not have to wait long for the right innovation.

Whereas chapter 1 shows that most of the innovations that life needs arise instantly, chapter 2 demonstrates something even more remarkable. Many new forms of life like grasses, ants, birds and mammals arose before their time had come, millions of years before. Some of them even went extinct again, not because they were flawed, but because the time was not quite right yet for them.

Ultimately, all innovations in biological evolution originated in the molecules that help build and maintain cells and bodies. Studying how nature changes these molecules is like peering into an inventor's brain. It can go beyond explaining *that* evolution innovates so prolifically, and help us understand *why* it does so.

Many innovations emerge only on this molecular level. They are entirely invisible, yet no less momentous than the creation of new species. Among them is photosynthesis, the ability to harvest energy from sunlight, discovered by cyanobacteria more than two billion years ago. Also among them is the ability to harvest energy from man-made toxins, a skill discovered by bacteria within the last century. These are two among thousands of innovations in metabolism. They go all the way back to life's origin, and allowed life to survive on myriad new foods and in myriad new environments. Many of these innovations are sleeping beauties. Chapter 3 explains why.

Especially important innovative molecules are proteins, the molecular workhorses that keep life going. Their tremendous innovative potential is revealed by a stunning 2008 medical discovery about Yanomami Indians from the jungles of southern Venezuela. These Indians had never been contacted by modern civilisation. Nonetheless, their skin and gut harboured bacteria resistant against not just one, but eight different modern antibiotics. These included first-generation antibiotics like penicillin, but also more recently developed antibiotics that serve as medicines of last resort against drug-resistant bacteria. Bacteria have a slumbering talent for resisting multiple drugs that they have never even encountered. This talent is a huge problem for medicine. It means that we may never win our race against antibiotic resistant germs.

Chapter 4 explains the origin of this remarkable talent, which endows proteins with hidden powers that go far beyond antibiotic resistance. Proteins that evolved to destroy one kind of toxin can destroy a dozen others, a talent that remains quiescent until the right kind of toxin comes along. In addition, proteins that synthesise one kind of defence chemical can also synthesise many others, a skill utterly useless until it becomes life-saving when the right enemy comes along. In other words, molecular innovations can be more than cheap: they can be free. Such innovations can help life invade new and hostile places, with consequences that can be devastating for humanity. They include multi-billion dollar economic damage caused by invasive species like water hyacinths. They also include suffering and death from multi-drug-resistant superbugs, all thanks to dormant innovations that awaken in the right environment. Conversely, we can also make such innovations work in our favour. They can help biotech engineers produce biofuels more efficiently, improve the enzymes that clean our laundry and use green chemistry to manufacture industrial chemicals.[2]

No less important than proteins are the genes in our genomes, because they encode every protein in our body. Each gene is a long sequence of DNA letters that encodes the amino acid string of one among thousands of proteins that keep us alive. Up to the early twenty-first century, biologists believed that new genes always originated when evolution modified old genes to endow them with new skills. But that belief turned out to be dead wrong. Exciting and revolutionary discoveries from the last ten years prove that evolving genomes incessantly create new genes from scratch, seemingly out of nowhere, from random strings of

DNA. Chapter 5 explains how and why. The reason is that creating new genes is much easier than we thought, because evolution can build on what is already there, not only DNA but a whole machinery of proteins that decode genetic information. In fact, in the cells of humans and other animals alive today, new genes are created all the time. But just like many other innovations, most of them perish quickly. Those that survive are often in for million-year waits, until their skills become useful in a changed world.

These and many other sleeping beauties show that no innovation succeeds on its own merit. The value of a new gene does not come from some inner quality of the gene. It comes from the world into which the gene is born, a world beyond the organism's control.

In contrast to natural innovations, whose substrates are genes and proteins, cultural innovations are made possible by sophisticated brains and their neural circuits. And in the realm of culture, sleeping beauties are just as abundant as in nature. The gateways to human culture are the tool-wielding animals of chapter 6, whose simple technologies are either hard-wired into an animal's brain or learned by an animal in its lifetime.

Dolphins protect their beaks with marine sponges when they dig in the ocean floor for hidden prey. New Caledonian crows bend twigs into hooks to scare insects from tree cavities. Monkeys use rock hammers and anvils to crack open nuts. Innovations like these are very different from those embodied in drug-resistant proteins, but they obey the same principles. An animal's ability to discover or use tools may be quiescent, until it is awakened in the right environment. And the success of any one tool innovation is not pre-ordained, but determined by the world it is born into.

The latent potential of animals to discover tools foreshadows a much more profound potential in humans. It is the potential to discover technologies as transformative as reading, writing and mathematics, technologies that enabled our modern civilisation. This potential is embodied in ancient human brain circuits that evolved long before humans did. Chapter 7 explains how their old job in other animals predestined them for new jobs in humans. Their potential lay dormant for many millennia. It was awakened by the agricultural revolution, which brought forth not just maths and writing, but myriad other manifestations of human culture. These include intricate musical instruments like pianos, sophisticated games like chess, and complex technologies like computers. Because the necessary skills have not been directly shaped by biological evolution, they highlight the immense, yes, nearly limitless, potential of a brain to use the old for new purposes.

Such a potential also exists on the most abstract level of thought, the kind that drove many human discoveries in science and technology. This is the subject of chapter 8, where I explain how our minds use analogies in the service of discovery. One such analogy compared atoms to vibrating strings, and helped early twentieth-century physicists understand phenomena like the radiation emitted by atoms. Analogies like this are much more than pedagogical tools. Likewise, their close cousins metaphors are more than just rhetorical tricks. Linguists such as George Lakoff discovered that they are foundations of abstract thought – they make scientific explanations possible. Any one such explanation is dormant, an unrealised potential, until a creative mind discovers the analogy leading to it.

Analogies and other new combinations of concepts are more than just peculiar forms of sleeping beauties. They help explain why our human culture overflows with innovations. We continually breathe new life into old, dormant things, including patterns of thought, by using them in new ways. This very ability is at the heart of human creativity in the arts, mathematics, sciences and technology, the focus of chapter 9, where sleeping beauties also abound.

Among them is the work of nineteenth-century poet John Keats, who by one measure was a complete failure as a poet. His volumes of poetry, taken together, sold no more than two hundred copies during his lifetime. This failure so embittered Keats that he asked that his grave remain nameless, adorned only with the inscription 'Here lies one whose name was writ in water.'

Keats really seemed destined to be forgotten. It would take decades before Keats' genius came to be recognised. That recognition began mid-century, when his first biography was published, and it culminated at the centennial of his death in 1921. By that time, Keats had become more than just a well-known Romantic poet. He had entered a pantheon of literary greats that included Homer, Shakespeare and Dante.

The list of creators neglected or ignored in their time is endless. It includes the painters Vincent van Gogh, Johannes Vermeer and El Greco, writers Emily Dickinson and Herman Melville, American naturalist and part-time hermit Henry David Thoreau, and even the composer Johann Sebastian Bach, whose music was considered old-fashioned and absurdly complex for more than half a century after his death.

Fickle fashion may help explain success or failure in the arts. Beauty, after all, is in the eye of the beholder. Truth, however, is not – at least in the hard sciences and in maths. But even maths and science are subject to fashions. Exhibit A is the development of linear algebra by nineteenth-century mathematician Hermann Grassmann. Today, linear algebra is central to university and even secondary school maths curricula. It is essential to solve linear equations. It is crucial for modern geometry, because it enables calculations with lines and planes in three and more dimensions. Engineering and science would be unthinkable without it. Yet when Grassmann developed many key concepts of linear algebra in an 1844 book entitled *Die Ausdehnungslehre*, he was widely ignored. He tried again in 1862 with a second edition, but had no greater success. Thirty years after the first edition, his publisher would write 'since your work hardly sold at all, roughly 600 copies were used in 1864 as waste paper.'

Even some fundamental laws of nature are ignored after their discovery. Among them is the law of energy conservation, first formulated by German physicist Julius Robert Mayer.[3] When others did not recognise its importance for years, Mayer attempted to commit suicide. (And when they finally did recognise its importance, they first gave credit to his British competitor Joule.)

As in science, so in technology. Among its sleeping beauties are some of the most profound innovations in human history. One of them is the wheel. This ancient technology first appeared independently in the Middle East and Eastern Europe some 3,500 years BCE, but its ascent was neither swift nor universal, as chapter 9 explains. Like many other inventions, humans discovered the wheel multiple times. Some inventions, like a cure for the life-threatening disease scurvy, are first ignored and then rediscovered, often multiple times, before finding success. They illustrate that revolutionary innovations appear almost inevitable from a broad historical perspective, although they may feel hard-won to individual inventors. But more important, they underline that success is guided by forces beyond the inventor's control.

From the molecular creations of nature to the technologies of humans, the examples of this book stretch over four billion years of innovation. Myriad innovations arise before their time has come. They lie in wait until that time arrives. No innovation, no matter how life-changing and transformative, prospers unless it finds a receptive environment. It needs to be born into the right time and place, or it will fail. The final chapter draws some lessons from these universal patterns of innovations for human creators, especially for those of us who are struggling, feel unrecognised, and may rail against being ignored or forgotten. Our creative process is a microcosm of a billion-year history of innovations. Knowing this deep history may not help us become more successful. It may not change our place in the world. But it can help us find this place and make the best of it.

PART ONE

Nature

1

Instant innovation

THE CATERPILLARS OF MONARCH BUTTERFLIES are addicted to
dangerous food. They devour the leaves of milkweeds, perennial
herbs that grow a few feet tall, with tiny star-shaped flowers that
form eye-catching inflorescences. Milkweeds may be beautiful,
but they are not innocuous. When the mouthparts of a cater-
pillar slice into a milkweed leaf, pressurised channels inside
the injured leaf release a substance that is milky white, hence
the name milkweed. The caterpillars know that this white stuff
means trouble, because they try to cut these channels and let the
milk drain before they devour the leaf.

The scientific name of this milk is latex, a complex and
sticky mixture of chemicals whose purpose is similar to that
of sticky resins, like the yellow excretions of pine trees. Such
excretions bring wound healing to mind, but what they do is
more sinister. That becomes clear from insects that were en-
trapped millions of years ago in the fossilised resin we call
amber. Latex and resins are lethal defensive weapons against
hungry animals.

When insects bite a chunk out of a latex- or resin-producing
plant, the sticky material can immobilise their mouthparts and
glue them together. It can even entrap them whole. More than
thirty percent of monarch butterfly caterpillars that hatch on the
leaves of milkweed get mired in the plant's latex, become glued

to the leaf, and die. And such entrapment is only one line of defence embodied in latex and resins. Both secretions can also contain toxic chemicals, such as cardiac glycosides, poisons that can quiet a beating heart forever. Latex and resins are sophisticated and complex chemical weapons.[1]

These chemical weapons are also examples of evolutionary innovations, so-called because they occur during the evolution of a species and help the species survive. And these specific innovations – here comes the important part – were not made by the milkweeds alone. Evolution discovered them not once, twice or a few times, but at least forty different times, on different branches of life's enormous tree, in completely different species. What is more, many of the species that discovered them also independently evolved a distribution network for these toxic excretions, an elaborate system of channels that deliver the sticky stuff wherever a plant is attacked.[2] Innovations like these, discovered multiple times, illustrate how easily evolution can innovate, and they foreshadow the sleeping beauties we will encounter later.

Latex, resins and their transportation networks are so important that biologists have elevated their status beyond that of mere innovations. We call them *key innovations*. That's not because they are hard to discover – their multiplicity shows otherwise. It's because they do more than just improve the survival of one plant species. They have more profound and long-term consequences for evolution.[3]

Latex-producing plants suffer less from insect damage, can grow faster, and spend more energy on reproducing by building flowers and producing seeds. These benefits allow latex-producing plants to spread further and colonise new habitats. In these habitats they eventually form new species. In the jargon of biology, such key innovations promote *adaptive radiation*, the sprouting of new branches on the tree of life. During an adaptive radiation, one species multiplies into many, each of them with its own lifestyle that is best suited to its own habitat.

The evolutionary potency of these chemical weapons is revealed by a study on sixteen plant families. Each family not only produces latex and resins, it has a closely related family that lacks this ability. The study showed that thirteen of the sixteen latex-producing families had evolved more species – not just a few more, but up to a hundred times more than the related families without this ability.[4] All in all, latex has been a smashing success in evolution. From its few dozen origins eventually emerged twenty thousand latex-producing flowering plants, as well as numerous latex-producing conifers, ferns and even fungi.[5]

Latex is an important innovation, but it is itself the product of an arms race that began with another innovation, an older one, discovered by insects. This innovation was phytophagy, the ability to use plants as food.

Hundreds of million years ago, primitive insects preyed only on other animals or lived on detritus, scavenging the carcasses of dead organisms. Switching from this diet to plants can't have been easy. One obstacle to the vegetarian lifestyle is the chemical warfare we just heard about, which plants excel at by necessity, because they can neither run nor hide. Another obstacle is that plant tissues and sap are poor in nutrients such as nitrogen and essential amino acids, much poorer than animal prey or detritus. What is more, detritus-feeders can hide in the ground, whereas plant-feeders take their meals out in the open. Their way of life exposes them not just to desiccation, but worse, to predators lured by their small, often poorly armoured and deliciously soft bodies.

Despite these obstacles, phytophagy was also discovered more than once, and not just two or three times, but at least fifty different times, by different species. And it became very successful, as biologist Charles Mitter from the University of Maryland and his collaborators proved. These researchers compared thirteen branches of plant-eating insects on the tree of life and thirteen nearby branches without. They showed that most of the plant-eating branches bear more species, and sometimes many more. For example, within an insect family also known as the green flies or *Chloropidae*, the plant-eating branch has more than 1,350 species, whereas the other has only 80 species. Even though only a few insect orders – large groups of species like beetles, cockroaches and dragonflies – discovered phytophagy, it became so successful that today, half of the world's 900,000 insect species feed on plants. Phytophagy is another one of evolution's key innovations.[6]

A plant-based diet poses yet another challenge: some plant-based foods like seeds or grasses are extremely tough. This challenge has been met in a unique way by yet another group of animals, the mammals. They evolved specialised teeth whose name – molars – comes from the Latin word for millstones, because they grind up plant parts.

If you reach into the back of your mouth to touch one molar – you have eight in total, or twelve if all four wisdom teeth have erupted – you will find them not entirely smooth but covered with small bumps, called cusps. Among these cusps is a key innovation of mammals, innocuous-looking but momentous. It's a bump called the hypocone, and you can find it at the rear end of a molar, on the side that is closer to the tongue than the cheek.[7]

The molars of primitive mammals had three cusps, and if you were to cut through one such molar horizontally, the cross-section would be triangular. That's a poor design for a tooth that is supposed to work like a millstone. The larger the surface of a grinding tooth is, the more plant material it can grind at any one time, but adjacent teeth with a triangular cross-section leave plenty of empty space between them, all of which is wasted for grinding. It would be much better if the cross-section were closer to a rectangle, and if molars were lined up in the mouth like bricks laid end to end. This is what the hypocone's evolution achieved. The addition of the fourth cusp filled in the missing space. It turned the cross-section from triangular to rectangular – the scientific term is quadrate – doubled the grinding surface, and allowed for more efficient grinding.

The hypocone evolved more than twenty different times on different branches of the mammalian tree of life. What is more, molars with hypocones became platforms for further innovations, because evolution embellished their basic architecture when it built mammals with specialised plant diets. Such mammals include deer, whose molars are strengthened by a crescent-shaped ridge of enamel. They also include elephants, whose molars bear multiple such ridges. These and many other molar shapes permit more efficient grinding, and all are derived from teeth with hypocones.

You may not be surprised to hear what happened next, when plant-eating mammals branched into multiple species, a radiation that started over fifty million years ago: those mammals with a hypocone, such as rodents and hoofed mammals, did better. They evolved the greatest number of species. Remarkable that, how a small chunk of tooth can alter the evolutionary paths of hundreds of species during millions of years.

One view of adaptive radiation is that a key innovation enables radiation, and that its absence prevents radiation. Only with a key innovation can a species exploit existing opportunities, such as a warmer climate, a new source of food, or a superior form of shelter. In this view, any one adaptive radiation has to wait, possibly for a long time, until the right innovation arises. And the need to wait holds evolution back. But innovations like latex production, phytophagy and the hypocone cast doubt on this view, because they have originated so many times.

Perhaps evolutionary innovation is easier than we think? Perhaps evolution does not have to wait for innovations? Perhaps its creative engine is more powerful than we give it credit for? The astonishing speed with which evolution can respond to new opportunities argues for this possibility.

This speed is showcased by plants best known for their striking blue or red columns of flowers, and their ability to fertilise the soil with nitrogen they harvest from the air. These plants are the *lupins*, legumes highly prized by gardeners because of their beauty. But they are more than just pretty. Their seeds – lupin beans – have been consumed by Europeans since antiquity. They have an even longer culinary history in the Andean mountains of South America, where people have been eating them for six thousand years.[8] And in the Andes the lupins have also experienced a remarkable burst of evolution.

Seven thousand kilometres long, the Andes are not only the world's longest mountain range, but also among the youngest mountains on the planet. Starting their rise some thirty million years ago, they did not reach their present topography until two to four million years ago. This topography includes one of the world's most peculiar habitats, the treeless northern highland known as the *páramo*, located mostly within Colombia, Venezuela and Peru.

Starting above the timberline at an altitude of three thousand metres, the páramo reaches up to the zone of eternal snow and ice beginning above five thousand metres. Despite its high altitude, the páramo's climate is clement, thanks to the nearby equator, even though it is chilly and temperatures can oscillate wildly. Moist air blown inland from the Pacific cools down as it rises to cross the Andes, and supplies abundant life-sustaining moisture in the form of rain, clouds and fog. Add to that the numerous habitats created by the valleys, slopes and crevices of a mountain range, and you begin to understand why the páramo is one of the world's biodiversity hotspots, with 45,000 species of plants, and forty percent among them endemic, living only there and nowhere else.

Among these endemic species are eighty-one species of lupins. They radiated from a single ancestor and evolved an astonishing diversity of forms and ways of life in the Andes. Some of them live for multiple years, others for one season only. Some stretch towards the sun with a stem, others hug the ground with a rosette of leaves. Some build only soft herbaceous tissues, others grow hardened and woody. The woody lupins themselves radiated into diverse forms. They include species whose branches lay low and creep close to the soil, species that grow into upright shrubs and species that develop into small trees.[9]

Even more remarkable is that these species emerged very rapidly from their common lupin ancestor, starting a bit less than two million years ago. In other words, not only did evolution bring forth eighty-one new species in less time than the five million years needed to create humans from an ape-like ancestor, it also created these species almost as soon as the Andes had risen and the páramo with them. Evolution closely tracked the rise of the Andes and rapidly exploited the new opportunities that arose with the páramo.

Other kinds of opportunities emerged from an altogether different geological process: the slow but inexorable fracturing of the African continent, a process so transformative that it will force us to rewrite our geography textbooks – eventually. It began some twenty-five million years ago when two plates of the continental crust began to pull apart in Eastern Africa. The process is called rifting and it produced not only towering volcanoes like Mount Kilimanjaro, but also the East African Rift valleys that extend over thousands of kilometres. These valleys partly filled up with giant lakes that may give way to a new ocean ten million years from now. Among them are lakes Victoria, Tanganyika and Malawi.

Lake Malawi and other East African lakes were colonised by numerous species of fish, among them the cichlids whose sixteen hundred species form one of the most diverse fish families. They include food fishes like tilapia, but also stunning aquarium fishes like the blue discus, whose flat, disk-like body (whence the name) is painted with what resembles a turquoise river network surrounding the orange island of its eye. It is only one of many scintillating gems in the treasure chest of cichlid diversity.

Lake Malawi's first cichlid colonisers arrived soon after the lake had formed some 4.5 million years ago, and when they did, they spread through the lake, but not only that. They started to radiate at astonishing speed, diversifying into some five hundred species at a rate that rivals the radiation of Andean lupins.[10] It may not be obvious how a lake could compete with the lifestyle opportunities of a richly textured mountain landscape – after all, it's just water – but plausibility can deceive. Some areas of a lakebed are sandy, whereas others are rocky, and different species prefer one or the other.[11] Also, a lake of 29,000 square kilometres – almost the size of Belgium – can support a variety of diets and feeding habits. Some fish strain plankton from lake water, others scrape algae off rocks. Some fish are indiscriminate and feed on a rich buffet of other animals. Others are picky and consume only one kind of food, such as snails.

Evolution's ingenuity becomes especially obvious from the more unusual Lake Malawi cichlids. Among them is one species that feeds only on the scales of other fish. Another such species is the sleeper cichlid, which plays dead until another unsuspecting fish ventures nearby. That's when it springs to life and devours the poor sucker.

Add to these feeding strategies the profusion of colours that cover cichlid bodies. One species is painted in bright signal red. Another species dons a bright yellow body with a solid black dorsal fin that resembles a mohawk on a punk rocker's head. Yet another sports cobalt blue stripes on a black background, bringing to mind a prison uniform. Such coloration not only helps potential mates find each other and avoid copulatory confusion. It also serves other purposes, among them the camouflage offered by vertical stripes, which are harder to see near a rocky lake bottom.[12] And all this diversity emerged nearly instantly in evolutionary time.

Lupins and cichlids are not alone as exhibits of evolution's agility. Other examples include the radiation of horses from a dog-sized ancestor, which began some twenty-five million years ago and closely tracked expanding grasslands that support a grazer's way of life.[13] They also include explosive radiations on islands or archipelagos like that of Hawaii, where 900 flowering plant species and more than 4,500 insect species emerged within ten million years.[14] Examples like these also point to another illuminating pattern of evolutionary change, this one revealed by the labours of palaeontologists.

Palaeontologists are skilled at reconstructing the anatomy of long-dead and bizarre creatures from fossilised bodies that are squashed, sheared, or shattered while the rocks that entomb them travel through the planetary crust. The skills of palaeontologists shine brightest when they are applied to rocks of different ages that are stacked like the layers of a cake. That's because the bodies reconstructed from such rocks can do more than just reveal the anatomy of an extinct species. They existed at different geological times, and can also help us understand how evolution transforms a body. Such a transformation was experienced by a peculiar species of fish some ten million years ago.

Three-spined sticklebacks are very different from the flashy Lake Malawi cichlids: they are comparatively drab, finger-long fish that inhabit oceans and lakes in the northern hemisphere. What's remarkable about them is an anatomical feature enshrined in their name. Their back is equipped with three flexible spines, which can either cling to the body or be raised so that they protrude like small daggers. What is more, where other species have two fins on their bellies, these sticklebacks instead sport two spines that can be drawn in or manoeuvred into a defensive position like the spines on their back. Add to these spines more than thirty hard plates of bony armour that cover their flanks, and it becomes clear that three-spined sticklebacks are no picnic for a would-be predator. Not only do their armour plates resist crushing, when those spines are erect feeding on a stickleback must be like trying to swallow a hedgehog.[15]

Sticklebacks have been around for millions of years, and during this time they have repeatedly left their oceanic home and colonised lakes, some of them created when glaciers melted away after an ice age. If a river connects such a lake to a nearby ocean, fish can colonise it by swimming upstream, but when the river dries up, their retreat is cut-off, and they are forced to survive in the lake. That need not be a problem, but when the lake itself goes dry, as most lakes do eventually, the colonisers are doomed to extinction.[16]

Some ten million years ago, three-spine sticklebacks initiated such a cycle of colonisation and eventual extinction from a long-ago shrivelled lake in present-day Nevada. Today the dry lake sediment is being mined for diatomaceous earth – fossilised algae useful as abrasives or insecticides – but it also harbours myriad fossilised sticklebacks. Because they died and fossilised at different times during the lake's existence, these fossils are a boon to palaeontologists, helping them monitor changes to stickleback anatomy over time and find out whether evolution altered this anatomy.

Indeed it did, as demonstrated by two studies that analysed five thousand stickleback fossils and tracked their changing anatomy. Sticklebacks that invaded the lake were highly armoured, but then they gradually lost their spines and their armour plating.[17] It is not hard to understand why. Building an armoured body with defensive spines requires materials and energy that are lost to other essential tasks like chasing food or making babies. The lake contained few predators, and if predators pose little threat, producing armour is wasteful. What is more, the raw materials needed for armour-plating and bony daggers are ions like calcium that can be scarce in lakes.[18] In other words, losing armour in this lake was the rational, the frugal thing to do for evolution.

What's remarkable is how rapidly evolution reduced stickleback armour. Rapidly on the time scale of evolution I should say, because it still took more time than most of us can imagine.[19]

We can easily grasp how the world changes during an amount of time similar to our life span, a bit less than a century. It is much harder to imagine how life has changed during the millennium since medieval knights battled each other, or during the two millennia since Caesar ruled the Roman Empire. Our imagination fails us for even longer time spans, like the ten thousand years since the beginnings of our current civilisation. Yet ten thousand years is about the time that these sticklebacks needed to adapt to lake life by losing their armour. And while being unfathomably long to us, ten thousand years is still incredibly short in evolution. It is less than one tenth of the time since modern humans originated, and less than one hundredth of the time that lupins needed to radiate in the Andes.

During this time, the number of spines and the strength of stickleback armour declined, rapidly at first, and then ever more slowly, until it was almost completely lost. We know this only because so many stickleback fossils have been preserved from this ancient lake that palaeontologists can resolve evolutionary time down to a few centuries. And that's highly unusual. Other ancient transformations of organisms are usually documented by many fewer fossils, and these fossils would be spaced much more than ten thousand years apart. In such transformations, ten thousand years would appear as a mere instant in time. In one moment the armour would be there, in the next gone.

Fossils do not document the evolution of most species so well, but the fossil evidence we have shows that stickleback evolution is typical: evolution responds lightning-fast to the challenges thrown at it. That's what palaeontologist Gene Hunt from the Smithsonian National Museum of Natural History demonstrated when he studied the evolution of more than 250 traits in more than 50 fossilised species ranging from single celled microbes to fish and mammals. When such traits evolve in a particular direction, like the declining armour in sticklebacks, they usually change rapidly. Most traits, however, do not change in any one direction, most of the time. Evolution's motto seems to be 'hurry up and wait'. [20] And when evolution needs to wait, it will often be waiting for geological upheavals, like those that created the Andes or Lake Malawi. Or it may wait for the kind of challenge posed by phytophagous insects, which renders latex and resins useful. Or it may wait for the tough plant foods that require quadrate teeth with hypocones.

The right kind of environment is essential for evolution, but it is not always sufficient. When cichlid fish colonised East African lakes, they radiated into multiple species in some lakes, such as Lake Malawi, but did not radiate in other lakes. Perhaps these lakes lacked the foods needed to support multiple lifestyles, or perhaps their colonisers had not experienced the right DNA mutations. [21] Unfortunately, we do not know. In fact, this ignorance is a long-standing problem for biologists, who have debated for many years whether the right starting species or the right environment is most important for evolution's advance. [22]

The answer might also help explain why some life forms have evolved only once. Among them is the African desert plant *Welwitschia mirabilis*. The plant is no beauty, resembling a wilted

lettuce with two gigantic leaves that lie on the ground and become increasingly shredded over the centuries of *Welwitschia*'s
long life. But the huge surface area of these leaves serves a purpose. It helps harvest moisture from fog that condenses on it
and drips to the ground.[23] Is *Welwitschia* unique because it is the
one best solution to survive in its unique desert environment,
where fog provides most moisture? Or is it a fluke of evolution, a
bizarre species that happened to survive, even though other species might have done better in this extreme environment? Might
they even have radiated? We have no idea, neither for *Welwitschia*, nor for other unusual species, among them the platypus,
the chameleon and the elephant.[24]

Fossils may not help us answer this question either. Not all
species fossilise well, and even when they do, the fossil record
is notoriously incomplete. It may mislead us into thinking that
a species is unique when it is not. Another problem is that our
entire planet is only a single gigantic experiment in evolution.

Fig. 1 *Welwitschia mirabilis*

Who knows what would happen if we could restart this experiment once, twice or many times. Palaeontologist Stephen Jay Gould asked this question in his book *Wonderful Life* and argued that life would come out very differently. But he really didn't know and neither do we. Who is to say whether a rerun would produce a completely different solution to *Welwitschia*'s problem, another *Welwitschia*, or no solution at all?[25]

A conundrum like this calls for approaching the problem in a completely different way. One such approach is pursued by scientists around the world – including scientists in my Zürich laboratory – who do not merely observe evolution in nature but let it proceed in a test tube. In other words, we and others perform laboratory evolution experiments. Such experiments are powerful, because they allow us to observe organisms evolving in real time, within years, months or even weeks. (Such experiments are also a potent antidote to the toxin of creationism.)

When we perform an evolution experiment, we expose an entire population of organisms to a new and challenging environment for many generations. The environment could be extremely hot or cold, wet or dry, it could harbour food that is nearly impossible to digest, intense amounts of damaging radiation, toxins like antibiotics or heavy metals and so on. If a population of organisms can survive in such a hostile environment at all, the DNA of some individuals will eventually experience mutations. Because these mutations are essential for evolution, their origins deserve some explanation.[26]

It's well known that DNA molecules form long strings called chromosomes that make up the genome of every organism. Each such DNA string is like a text written in a chemical alphabet consisting of four letters – four small building blocks called nucleotides and commonly abbreviated as G for guanine, A for adenine, C for cytosine and T for thymine. Each chromosome string harbours hundreds to thousands of genes, shorter stretches of DNA that encode the information needed to build an organism. This information is decoded by first copying or *transcribing* the gene into RNA – a molecule with building blocks very similar to DNA – and then *translating* the RNA into protein, another molecular string, but with very different building blocks called amino acids, twenty different kinds of them.

A protein string is highly flexible, and in the chaotic environment of a cell, where heat causes atoms and molecules to vibrate and shake incessantly, millions of jittery molecules bump into this string, causing it to wiggle and to jiggle. Driven by all these bumps, a protein folds into a three-dimensional shape. This shape – also called a protein's fold – can resist ongoing molecular bounces, because some of its amino acids attract and stick to each other when they are near each other in the fold. Their stickiness stabilises the folded protein. A cell contains thousands of different kinds of proteins, each decoded – biochemists say *expressed* – from the information in a gene, each with a different fold. And this fold endows each protein with a special skill. Transport proteins import nutrients or export waste, cytoskeletal proteins give shape to cells, signalling proteins convey messages between cells, enzymes catalyse chemical reactions and so on.

Each cell of each organism on this planet is a hub of frenetic protein activity, where thousands of proteins simultaneously import, build, aggregate, cleave, export or destroy other molecules. This frenzied dance of molecules sustains life, but it also has unintended consequences. For example, when enzymes harvest energy, they sometimes create highly reactive waste products called free radicals that will damage or destroy other nearby molecules when they crash into them. When that happens to DNA, the DNA becomes damaged. Cells employ specialised protein machines that labour tirelessly to repair such damage, but unlike the Pope, these machines are not infallible. Sometimes they screw up. For example, when finding a mutilated letter, such as a damaged A, they will not repair it, but change it to a G, C or T. That's how an important kind of DNA mutation is born. It is called a point mutation, because it affects the smallest part of a genome, a single DNA letter.

Cells use an equally fallible protein machinery to copy their DNA before they divide, to make sure that every mother cell can pass a copy of its genome to its daughter. And when this machine makes copying errors, it can also create point mutations. Yet other point mutations occur when UV light, X-rays or other forms of high energy radiation careen into DNA, and permanently alter its letter sequence.

Some of these mutations may not affect an organism's survival, but most others will sicken or kill it, because they break some part of the molecular protein machinery that sustains life. Yet other mutations have the opposite effect. They are beneficial mutations, creating or improving the skills an organism needs to survive in a new environment. They are the rarest mutations, but also the most interesting and important ones. They help create proteins that can harvest energy from novel foods, destroy and disarm potent toxins, or protect life against excessive heat or cold.

During an evolution experiment, any one individual in an evolving population can be hit by such a beneficial mutation. And when that happens, the individual will either reproduce faster or its chances to survive will increase. Over multiple generations, its descendants will therefore outcompete other individuals. There will be more and more of these descendants, and they will experience additional mutations, most of them damaging, but a few of them beneficial, which improve the descendants further, and cause *their* descendants to spread through the population. As time passes, generation after generation, this interplay of mutations and natural selection will steadily improve survival, helping the population's members to adapt to their environment. And it will cause more and more beneficial muta-tions to accumulate in the survivors' genome.

During experimental evolution, organisms may begin to change after fewer than ten generations, but the longer you wait, the more obvious their changes become. In practice, it's best to let evolution run its course for at least a hundred generations. That could be a very long time for large organisms like some mammals, whose generations are measured in months or years. Fortunately, many smaller organisms reproduce much faster. Among them is *Drosophila melanogaster*, the tiny fly that feasts on rotting fruit in our kitchens. A single *Drosophila* generation lasts about two weeks, and one year of experimental fly evolu-tion would cover twenty-four generations. The longest running evolution experiment on this fly has been going on for more than three decades. It has covered more than eight hundred fruit fly generations, corresponding to some twenty thousand years of human evolution.[27]

These numbers sound impressive but they pale beneath the number of generations we can comfortably observe in a bacte-rium like *Escherichia coli*, which divides multiple times a day. In such a bacterium, an evolution experiment of a thousand generations need not last longer than a few months. The lon-gest-lasting such experiment with *E. coli* is still ongoing. It was started in 1988 by Michigan State University biologist Richard Lenski, and has been running for more than seventy thou-sand generations, the equivalent of 1.5 million years of human evolution.[28]

Numbers like these also highlight the limits of experimen-tal evolution. We may never recreate spectacular radiations like those of cichlids or lupins in the lab. That's because evolving large organisms for enough generations would simply take too much time. Fortunately, adaptive radiations are not the only measure of evolutionary success. Evolving the ability to conquer

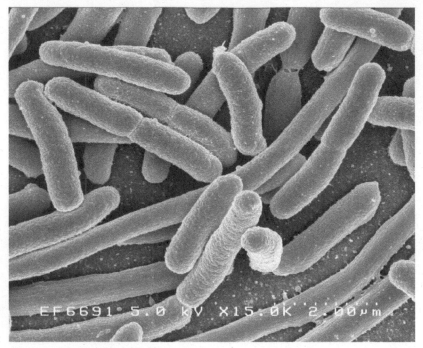

Fig. 2 *E. coli* under the microscope

and thrive in an extremely hot, dry, toxic or otherwise challenging environment can be just as important.

Experimental evolution also offers other tremendous advantages over observing evolution in the wild. First, it helps us understand how the environment affects evolution, because we can control this environment in the lab with exquisite exactitude. We can adjust the temperature to a tenth of a degree, and gradually ramp it up, challenging an organism to evolve heat tolerance. We can poison the organism with antibiotics dosed to milligram accuracy, and goad it to evolve drug resistance. And we can leave out an essential nutrient from the organism's diet, spurring it to innovate around this problem and evolve ways to synthesise the nutrient or survive without it.

Even more important, experimental evolution allows *replication*, a technical term for performing multiple identical experiments in parallel and comparing their outcome. Lenski's team of researchers, for example, has been evolving not one but twelve *E. coli* populations, each in the same environment and starting from the same *E. coli* population. Replication is crucial, because it allows us to find out whether evolutionary success is rare and unique, or frequent and repeatable.

That's precisely the question we cannot answer for apparently unique species with an incomplete fossil record. Fortunately, many experiments in laboratories around the world have answered this question for various species evolving in different environments. Researchers in my laboratory, for example, have repeatedly evolved microbes in environments containing toxins like antibiotics, amounts of salt that would be lethal to humans or chemicals that release DNA-damaging free radicals.[29] And we find that microbes generally adapt rapidly to their new environment, increasing their ability to grow and survive to a similar extent in replicate experiments. The work of many other laboratories conveys the same message: evolution helps populations adapt quickly to their environment.

But the proverbial exceptions that confirm the rule also exist in evolution experiments.[30] Evolution sometimes fails, at least temporarily. One instructive example involves Lenski's 70,000-plus-generations evolution experiment with *E. coli* populations.[31] The environment of these populations contained the sugar glucose as one source of energy, and in all twelve parallel experiments, *E. coli* evolved to use this sugar more and more efficiently, growing and dividing faster and faster as the experiment went on. But the environment also contained small amounts of citrate, another potential source of energy. Unfortunately for *E. coli*, the bacterium is not able to use citrate, because it is not able to import it into the cell. For the first 31,000 generations, it looked as if things would remain that way, because none of the twelve evolving *E. coli* populations could survive on citrate – failure. But after 31,000 generations, one

E. coli population experienced mutations that modified a transporter protein and enabled it to import citrate.[32]

To put such unique mutations in perspective, keep in mind that even though bacteria reproduce rapidly, and even though that experiment lasted decades, those decades are still a short amount of time in bacterial evolution. Continue that experiment for a century or a millennium – even *that* is much shorter than the time that sticklebacks needed to scale back their armour – and most populations might yet experience mutations allowing them to use citrate. In other words, apparently unique innovations may become common as time goes on. An apparent failure to adapt may thus be even rarer than experimental evolution lets on. Other laboratory experiments from the 1980s underscore this point. They had discovered similar citrate-enabling mutations independently from Lenski's bacteria and long before it.[33]

Experimental evolution allows us to accomplish more than just observe evolution in action. At the end of an evolution experiment, we can also use the technology bestowed on us by twentieth-century science to read – to *sequence* – the DNA of the survivors and to study the innovative mutations that helped them survive. And when we do, we find even more evidence for evolution's agility: the beneficial mutations of well-adapted survivors often do not occur in the same genes. Different survivors harbour mutations in different genes, sometimes dozens of different genes. In other words, different solutions to the same problem exist and evolution can discover them, even during a short laboratory evolution experiment. For example, when bacteria evolve to survive lethal antibiotics, mutations affect three different kinds of proteins. The first are the very proteins that an antibiotic attacks – mutations can immunise these proteins against the attack. The second are proteins that can pump antibiotics out of the cell – mutations can help them pump faster. The third are proteins that cleave and destroy antibiotics – mutations can help them become more efficient. Each of these are different solutions to the same problem.[34] Each of them arises multiple times in different replicated experiments.

In sum, experimental evolution reinforces what we learned from nature's grandiose experiments, those that brought forth key innovations like the hypocone and latex. With few exceptions, evolution can react instantly to the challenge of a new environment in the laboratory, just as it does in the wild, where it rapidly responds to changes like the rising of a mountain range. It rapidly discovers solutions to the problems we throw at it. More important, it discovers each solution not just once but multiple times. It may even discover multiple different solutions, and each of them more than once. All this underscores that innovation comes easily to biological evolution.

The examples from this chapter leave us with two possibilities for evolution's potential to create. The first is that evolution is no more than a nimble innovator. It reacts just-in-time to a changing environment and tracks many such changes closely over time. If so, it innovates only in the wake of evolutionary opportunities, like the rise of a mountain range or the rifting of a continent. The second possibility is more tantalising. What if many innovations arise *before* their time? If so, they would remain dormant, sleeping beauties that would awaken only when the conditions for success are right. In this case, evolution would be much more than just a nimble innovator. Indeed, as we shall see in the next chapters, nature often innovates before the need arises – long before.

2

The long fuse

WHEN THE NOMAD HORDES OF Genghis Khan spilled out of the Mongolian steppes to conquer Asia, Europe and the Middle East, their battle tactics gave them an edge over settled civilisations with lumbering armies.[1] Avoiding direct confrontation with technologically sophisticated enemies (think: knights in shining armour), the Mongols perfected the art of feinting retreats, luring their enemies into pursuit, and then picking them off from afar with bow and arrow.

Their supremacy goes back to a single cause: grass.

Grasslands are harsh environments: they see too little rain to sustain cool and shady forests, they experience dry seasons where life barely scrapes by, and they do not support enough vegetation to permit permanent human settlements. For these reasons, grasslands force humans into a nomadic lifestyle to gather enough food, hunt enough animals or raise enough livestock to survive. And where settlements are a liability, survival comes down to mobility. It was this mobility that helped create the nimble battle tactics of steppe warriors. It also prepared these warriors to move equipment and troops through thousands of miles of the Eurasian steppe, which was like a superhighway for horse-bound people.

Long before grasslands provided the stage for this clash of peoples, they already played an even more important role in human civilisation: they helped create it. Grasses were central to the origins of agriculture some ten thousand years ago. I use the plural 'origins' for a reason, because agriculture originated more than once, a bit like the evolutionary innovations discussed in the previous chapter. In fact, agriculture originated at least eleven different times, in distinct regions including the Middle East, China, Africa, New Guinea and Middle America, and with crops including wheat, rice, sorghum, sugarcane and corn. All of these crops are grasses. To this day, grasses make up seventy percent of human crops and provide more than half of all the calories we consume.

But even before humanity settled into villages and cities – probably long before humanity itself arose – grasses already played an outsized role in the biosphere. Just consider that they shape the life cycle of many animals. For millennia, bison have roamed the North American prairies in search of greener pastures, wandering north in the spring and south in the winter. And in the savannahs of Tanzania, millions of wildebeest have been migrating annually for ages, tracking a green wave of grasses that follows the rains.[2]

Grasses cover huge parts of the world's continents, from African savannahs to South American pampas, and they can even thrive in ecosystems like forests and wetlands where they do not dominate.[3] By at least two measures of success – their sheer abundance, and their diversity of ten thousand species and lifestyles – grasses are a triumph of evolution.

That's why it is so surprising that for most of their evolutionary history – many million years – grasses didn't look destined for a great future. They merely subsisted at the biosphere's margins.

We know from several sources that grasses originated in the age of dinosaurs more than sixty-five million years ago. The first source, odd yet reliable, is fossilised dinosaur dung. It contains phytolites, small and hard grains of silicon dioxide that many plants deposit in their tissues. Phytolites are mineral finger-prints of plant species, because their shape and size differ among species. The grass phytolites in dinosaur dung tell us that grasses are ancient. (They also tell us that a grazing way of life is as old as the dinosaurs.)[4]

Phytolites agree with a second source of information about grass origins: pollen. Just as grasses produce signature phyto-lites, they also produce signature pollen, and such pollen already appears in fossils that are more than sixty-five million years old.[5]

But unfortunately, if we want to find out when grasses – or any other kind of organism – first appeared in evolution, we cannot just rely on fossils. Dead bodies become fossilised only under special conditions, for example when they fall to the bot-tom of a lake or ocean, and quickly become covered in mud or sand, which allows them to become mineralised and turn into rock. When a new kind of organism first appears in evolution, it exists in small numbers, reducing the chances of that happen-ing. And even when it happens, we may overlook a new and rare fossil among thousands of other, more numerous fossils of abundant species.

Fortunately, a nifty tool helps biologists to measure time even where fossils are scarce. We can take advantage of numer-ous clocks that are slowly ticking away inside each cell of each organism of each species on earth. These molecular clocks are passed on from one generation to the next, and they record un-fathomably large amounts of time – eons, really – from millions to hundreds of million years.

Let me explain. I already talked about point mutations – changes in single DNA letters – that occasionally hit an or-ganism's DNA. When a gene suffers such a point mutation, the mutated DNA letter sequence is transcribed into a mutated RNA sequence, which can result in a protein where a single amino acid has been replaced with another amino acid. Such a change usually does more harm than good. It renders protein enzymes more sluggish, transport proteins less efficient, or cytoskeletal proteins less rigid. In the worst case, it can inactivate the mutated protein – knocking it out, as geneticists like to say – and when that protein is essential for life, the organism will die. But not all point mutations are harmful. Some do not affect a protein at all, because the protein can tolerate them. Such changes are called neutral mutations. A neutral mutation is invisible to natural selection, and can thus be passed on easily from one generation to the next. When that happens, a molecular clock has ticked once.

Each gene is like a clock that ticks whenever one of its DNA letters changes, and whenever that mutation is not vanquished by natural selection, but passed on from one generation to the next. The odds that any one DNA letter mutates are tiny, somewhere between one in a million and one in a billion during the lifetime of an organism. But because a gene can encompass thousands of letters, and because organisms pass on their DNA for thousands and millions of generations, these odds add up. And they can be measured by experiments which count the mutations that accumulated over time in different genes and genomes.[6]

Genomes harbour thousands of genes, and each of these genes is a molecular clock, but not all of these clocks tick at the same rate. Some genes encode proteins that can tolerate many mutations. Mutations in these genes frequently survive. Other genes encode proteins that tolerate few mutations. Mutations in these clocks rarely survive. In other words, some clocks tick rapidly, a bit like stopwatches that measure time in milliseconds, whereas others tick slowly, like the clocks of a church tower that measures time in minutes and hours. The analogy is imperfect, though, because molecular clocks tick much more slowly than human clocks. Fast molecular clocks measure the passing of millions of years, slower ones the passing of tens of millions of years, and the slowest ones measure time in hundreds of millions of years.

One complication in using these clocks is that not all organisms harbour the same clocks, because not all genomes harbour the same genes. The reason is that different organisms live in different environments and need different genes to survive. For example, because the gut and the soil contain different nutrients, gut bacteria and soil bacteria need different protein enzymes – all encoded by genes – to digest these nutrients. A gene that may be essential to survive in the gut may be useless to survive in the soil. And when a gene is useless, it will eventually disappear from a genome through knock-out mutations. When that happens, an organism has also lost one of its molecular clocks.

The most useful molecular clocks are genes that most organisms need to survive. They are also called housekeeping genes, because they perform essential jobs in a cell. They include genes that help copy or repair DNA, genes that help transcribe and translate DNA, and genes that help store energy. One of them encodes a protein called ATP synthase, which synthesises an energy-rich molecule called adenosine triphosphate or ATP that all organisms – including grasses – use to store chemical energy. Another one encodes a plant protein called RuBisCO, an acronym that stands for – get ready for this – ribulose 1,5 bisphosphate carboxylase oxygenase.[7] The name is as daunting as RuBisCO is crucial for photosynthesis. One hint of its importance: RuBisCO is the most abundant enzyme on the planet – among all proteins that all organisms in the biosphere produce, they produce RuBisCO in the largest quantities.

RuBisCO's job is to take a carbon dioxide molecule from the air and attach it to a sugar molecule. That doesn't sound impressive, but it's a key step in a longer sequence of chemical reactions that transform atmospheric carbon into building materials for plant cells, and ultimately into all plant tissues. Each atom of the 450 billion tons of carbon stored in the plants of our planet has passed through a RuBisCO molecule.[8] Crucial indeed.

Imagine that a molecular clock like RuBisCO ticks about once every million years. And imagine two species – call them X and Y – that harbour this clock gene. Suppose also that the gene's DNA differs in ten letters between the species (Figure 3.1). This difference tells us that the species' common ancestor – more accurately, their most recent common ancestor – lived approximately five million years ago. That's because the species evolved separately since then, such that about five of the ten mutations have accumulated along the branch leading to X, and the other five on the branch leading to Y (Figure 3.1). In other words, the same molecular clock ticking away in each species would have ticked five times along each branch.[9] The same idea applies when you add a third species Z. If, say, the clock gene differed in twenty letters between Y and Z, you could infer that Y and Z went their separate evolutionary ways some ten million years ago (Figure 3.2).

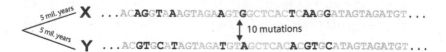

Fig. 3.1

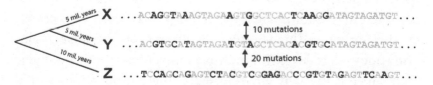

Fig. 3.2

Think of the three species in figure 3.2 as three leaves on a tiny branch of the vast tree of life, just like you can think of all groups of related species – grasses, lupins, butterflies – as forming a single branch of this tree.[10] Tracing the evolutionary history of any one species is like a journey on this tree, a journey back in time. It is like travelling backwards from one leaf towards the trunk, beginning with the small branch from which the leaf sprouted, to the larger branch from which the smaller branch emerged, and from there to the next larger branch, and so on. The place where two branches join into a bigger branch corresponds to a special moment in evolutionary time, the moment when the most recent common ancestor of species on the two branches existed. And even though the tree of life is much larger than any tree on earth, with millions of species and just as many branches, we can find that moment in time. We simply need to count how often a molecular clock ticked along the journey back from the leaves to the fork – the common ancestor – from where their shared branch grew.

When biologists do that for multiple species of grasses, using for good measure not just one but two molecular clocks – ATP synthase and RuBisCO – they find that these species shared a common ancestor at least sixty-five million years ago. In other words, molecular clocks independently confirm what fossils tell us: grasses are an ancient plant family.[11]

That ancient origin itself is not surprising. Grasses may be old, but many other forms of life are older. More surprising is what happened after their origin: not much. Grasses did not become today's dominant species until forty million years after their origin.[12] This is especially mysterious, because evolution endowed grasses with multiple innovations that would give them an edge in many environments. These innovations would shine especially in harsh and dry environments, which also happen to be replete with the number one enemies of grasses: grazing animals.[13]

One of these innovations altered the basic architecture of a plant's body. In most plants, cells above ground divide only at the tips of shoots, in places called *meristems*. Those are also the only places where a plant grows. But this feature creates a deadly problem when confronted with a grazing animal: once the grazer has eaten those vital shoot tips, the plant will die, because it can no longer grow and regenerate. In grasses, evolution solved this problem in a simple yet ingenious way. It moved the meristems away from the shoot tips and close to the ground. Even when a grazer leaves only a stump of a grass blade behind, the blade can regrow from this stump. What is more, the tender and moist tissue harbouring these dividing cells is enclosed by a leaf that grows like a sheath around it, which helps grasses survive drought by preventing water from evaporating.

Less visible but just as important are the chemical defences of grasses. These include lignin, the same substance that makes wood tough. They also include silicon dioxide, which is the hard stuff that sand and phytolites are made of. The blades of grass are rich in both, which makes them difficult to digest. In addition, silicon dioxide also wears down teeth and can cause kidney stones. Grazing animals may not worry about kidney stones, but they dislike having their teeth sanded down to stumps. Given a choice, they prefer the tender leaves of trees.

The benefits of lignin and silicon dioxide don't end there. They also protect against drought, because they stiffen leaves and prevent them from wilting. Yet another drought-protecting feature is the root, which can store nutrients and help a grass regrow quickly when the rains come back.[14]

All of these are important innovations, but perhaps the most profound is another chemical one. It is called C_4 photosynthesis.[15] The name comes from a molecule with four carbon atoms that some plants produce when they harvest carbon dioxide from the air during photosynthesis. (These plants are also called C_4 plants. In contrast, more conventional C_3 plants produce a molecule with only three carbon atoms.)

C_4 photosynthesis mitigates an important flaw in RuBisCO, the enzyme that harvests carbon dioxide from the air. When RuBisCO does its job, the enzyme sometimes mistakes oxygen for carbon dioxide, and attempts to incorporate oxygen into plant biomass. That's a problem, because oxygen is a waste product of photosynthesis. It's as if a cook left out one essential ingredient of a pie, but mistakenly added a dollop of garbage to the dough. We humans would throw out the entire pie, but plants are less wasteful. They remove and recycle the wrong molecules RuBisCO helped build. But when they do, they pay a price in life's universal currency: energy.

This cost is small when the air around a plant contains little oxygen and a lot of carbon dioxide. In that case, RuBisCO usually grabs the right molecule. But the cost can become enormous when the air contains a lot of oxygen and little carbon dioxide. That's when RuBisCO often grabs the wrong molecule, and every time that happens, it costs the plant energy. This problem arises especially in hot and dry places. The reason is simple. When it's hot and dry plants close the tiny pores in their leaves – stomata – through which they take up fresh air, because they would otherwise lose too much water. But when they hold their breath in this way, no new carbon dioxide can enter a leaf, and oxygen waste builds up, a bit like when an aquarium fish poisons itself by peeing into its bowl. In other words, plants in hot and dry places face a tough choice: lose precious water and preserve energy, or stay moist and waste energy.

C_4 photosynthesis cuts this Gordian knot with an ingenious sequence of chemical reactions. Their purpose is to concentrate carbon dioxide around RuBisCO, thus reducing the risk that RuBisCO grabs the wrong molecule. C_4 plants can have it both ways. They can preserve water *and* save energy.[16]

C_4 photosynthesis is remarkable for yet another reason. It originated at least seventeen times independently in the evolution of grasses.[17] We know this, because the major grass-bearing branch of life's tree can be subdivided into multiple smaller branches, twelve subfamilies that fall into more than seven hundred genera, which are units of biological classification below the family and above the species. Only some of these smaller branches hold grass species with C_4 photosynthesis, and – this is the important part – they are separated from each other by branches that do not. Their separation tells us that each of them had an ancestor that discovered C_4 photosynthesis on its own.[18]

With innovations like low-lying meristems, grazing defences and C_4 photosynthesis, you'd think that grasses would do very well. The fact that they didn't – for the unimaginably long time of forty million years – tells us that evolution's failure to innovate was not at fault. And when grasses finally exploded in numbers, evolution had just as little to do with it, because it did not create any further spectacular grass innovations. Instead, the success of grasses came from somewhere else: the planet's climate.

Since the earth accreted from stellar dust, its climate has been ever-changing. The heat of the sun has been waxing and waning, the earth's elliptical orbit has been oscillating ever so lightly, the earth's axis has been wobbling like that of a spinning top, mountain ranges have been rising from crashing continents, ocean currents have been wavering and swaying, meteorites have been striking and volcanoes erupting, exhaling greenhouse gases or darkening the sky. Some of these causes have been acting over millennia, others over years, but all have been conspiring to create a climate whose only constant is change. Over the past four billion years, they have created climate extremes like that of a hothouse earth, with little or no ice on the poles and sea levels two hundred metres higher than today. They also created frigid ice ages that buried large parts of continents under mile-high ice sheets.

Among all these changes was the one that triggered the rise of grasses. It was a drying of the earth in the middle of the Miocene, an epoch of earth history that started twenty-three million years ago. This change was a momentous climatic shift, caused in part by the drifting and shifting of continents, which created highlands like the East African plateau, as well as mountain ranges like the Sierra Nevada in California and the Andes in South America. Such mountain ranges create an enormous rain-shadow. This means that passing rain clouds, blown in one direction by prevailing winds, dump most of their water on one side of the range, leaving little for the other side. The results are vast arid regions like the Atacama Desert in the rain shadow of the Andes, or the Great Basin spanning Nevada, Utah and Oregon, which lies in the rain shadow of the Sierra Nevada.

In this period of climate change, millions of forested square kilometres no longer received enough water for a forest. And when the forests retreated, the grasses were ready, their innovations finally paying off. At about the same time, the amount of carbon dioxide in the atmosphere fortuitously declined, giving C_4 plants another boost, because they do better than other plants when carbon dioxide is scarce.[19] The result was that grasses expanded hugely in numbers until they covered habitats as different as the African savannah, where grasses are interspersed with scattered, drought-resistant trees, and the vast treeless steppes of Eurasia, from which the Mongolian hordes would erupt millions of years later.

In sum, all their innovations did grasses little good until the right planetary climate came along, which took millions of years.

Grasses were not the only plant family that had to wait this long for success. Another one innovated its way to even greater drought and heat resistance than grasses, and with distinctive but no less ingenious solutions. These are the cacti. Whereas evolution equipped the canopies of trees with an enormous surface area to absorb sunlight, in cacti it pulled all that surface inward to preserve water, transforming their bodies into spherical or columnar shapes. It converted their stems into water storage organs, covered them with moisture-retaining waxes and wind-stopping hairs that reduce evaporation. It also furnished them with shallow roots, so they can pick up whatever little moisture hits the soil – even dewdrops can be enough. It also endowed them with C_4 photosynthesis, and even helped them pull off another water-preserving trick: cacti open their stomata to take up and store carbon dioxide only at night. They consume this carbon dioxide only during the day, when they keep their stomata closed to preserve water.

Not quite as old as grasses, cacti originated in the Americas some thirty-five million years ago but did not leave much of an impression on the world until twenty-five million years later. That's when parts of the American continents became especially dry, drier even than grasslands, forming forbidding wastelands like Mexico's Sonoran Desert. And that's when cacti exploded in diversity. Within a few million years they radiated into the eighteen hundred species alive today.[20]

* * *

All my examples so far involve plants, which are especially helpless when coping with a changing world. That's because they are rooted to the spot. They cannot simply relocate and choose a better environment, one where their survival skills can help them prosper. You may suspect that this is why so many plant species experience delayed success. Perhaps only they had to wait?

That may be plausible, but it is also wrong. Some animals had to wait too, and even longer. Among them are the hairy, live-bearing, warm-blooded, lactating animals we are all familiar with, because they are us: mammals.

Just like grasses and cacti, the 5,400 mammalian species evolved astonishingly diverse forms. Among them is the colossal blue whale, whose weight can exceed 150 tons. That's just an estimate and it may even be low, because weighing blue whales is hard. (It also involves a gory procedure: slicing a hunted whale into sections about half a metre wide, and weighing these sections separately.)[21]

Contrast that behemoth with the smallest known mammal, the Etruscan shrew, with a mass as small as 1.3g, less than a hundred millionth of the whale's mass. The shrew makes up in cuteness what it lacks in size, resembling a tiny mouse with a long snout. That resemblance is superficial, though, because shrews are not rodents but relatives of moles and hedgehogs that feed on insects. The Etruscan shrew does so with particular abandon, devouring up to twice its body weight in food per day. It needs all this food to keep its hyperactive warm-blooded metabolism going.

Shrews also match the textbook stereotype of those early mammals that already lived in the age of the dinosaurs. Nimble and small, these animals ran circles around the lumbering dinosaurs. Their warm-blooded bodies and furry insulation allowed them to be active at night, the better to avoid getting eaten. And when the dinosaurs finally perished in the climatic cataclysm triggered by a giant asteroid that crashed into the Yucatán peninsula, the mammals rose from underneath their vanquished oppressors and conquered the continents. Or so the story goes.

There is some truth to this stereotype. Most early mammals were small. Some did resemble shrews. They did coexist with dinosaurs. And they did not explode in numbers, increase in size, or evolve today's diversity until the dinosaurs went extinct sixty-five million years ago at the end of the Cretaceous period. But the stereotype also ignores spectacular paleontological discoveries from the last thirty years, embodied in complete mammalian skeletons discovered mostly in China.[22]

These skeletons teach us that mammals are much older than previously thought.[23] For example in 2002, a Chinese–US team of scientists discovered a sensational fossil of a new mammalian species they called *Eomaia scansoria*.[24] *Eomaia* was some ten centimetres long, and it is so well preserved that even remnants of its fur are clearly visible. Until its discovery, the earliest mammals were thought to be seventy-five to eighty-five million years old. That date is already remarkable, because it's ten to twenty million years before the age of mammals began. But *Eomaia* pushed mammalian origins back by forty to fifty million years, because it is 125 million years old.[25]

Eomaia is Greek for 'dawn mother' – a mother of mammals – and *scansoria* comes from the Latin word *scandere*, to climb. The name gives away that this animal was no simple shrew, even though it was small, about ten centimetres long. Its elongate limbs and stretched digits are that of a climbing animal. In fact, these limbs resemble those of today's tree-climbing monkeys more than those of tree shrews.[26] In other words, *Eomaia* was an experiment in making a living on a tree. Sadly, it left no descendants, and went extinct again long before the end of the Cretaceous.

Eomaia was not the only, nor even the earliest mammalian experiment with life on trees. Tens of million years before it, deep in the Jurassic era lived another mammal-like animal called *Henkelotherium*. It was another shrew-like tree dweller that went extinct long before the age of mammals arrived. Yet another extinct tree-living mammal is *Juramaia*, which dates back to 160 million years ago.[27]

Mammals arose almost a hundred million years before dinosaurs went extinct. During all this time, mammals remained the proverbial underdogs, but mammalian evolution did not stand still. It experimented with various mammalian lifestyles, and a tree-living lifestyle was only one of them. Another was a water-living lifestyle. By 160 million years ago, evolution had created another hairy mammalian ancestor called *Castorocauda*, which had webbed feet, a dead giveaway for an animal living in water. It also had teeth specialised for grabbing fish and a flat, scaly tail. In other words, it resembled a beaver, a small one though, because it weighed less than a kilogram. However, that resemblance does not indicate true kinship with today's beavers, because *Castorocauda* also went extinct long before modern beavers arose.[28]

Volaticotherium antiquus – Latin and Greek for 'ancient flying beast' – was yet another ancient mammal with a unique lifestyle. It sported grasping toes to hold on to tree branches, a long flat tail for balance and most revealingly, a wingsuit-like membrane that connected its forelimbs and hindlimbs. This membrane is called a patagium, and is also found in today's tree squirrels. In other words, *Volaticotherium* was a gliding mammal that hurled itself from tree to tree, although its teeth are not like those of squirrels but more like those of an insect eater.[29] It too went extinct.

And there was *Fruitafossor*, discovered near the city of Fruita in Colorado. It had powerful forelimbs to dig into hard soil, teeth specialised for eating insects like termites, and resembled today's anteaters or armadillos.[30] Anteaters, sadly, cannot claim it as an ancestor, because its lineage also died out.

Long-extinct experiments like these illustrate that mammalian evolution repeatedly solved the same problem – how to survive on a tree, in water and in air. It also came up with solutions similar to those embodied in today's mammals. This phenomenon – different organisms, similar innovations – is so widespread it has its own name: convergent evolution. What is more, it tells us something about the shape of the mammalian branch of life's tree.

While this branch grew, a few side branches sprouted from it, evolutionary experiments that discovered and rediscovered the means of survival embodied in tree shrews, beavers, flying squirrels and anteaters. And although these side branches eventually shrivelled and died, the main branch somehow muddled its way through the age of dinosaurs. It continued to grow and eventually erupted into a riot of new species. Because it did not immediately sprout the five thousand-plus leaves of today's mammals, it resembles a rose that grows a long stem before it bursts into a riot of petals.

The trigger for this burst was not any one innovation, because the mammalian innovations had been made before – many of them more than once. The trigger was once again a change in the world. One such change was the dinosaur extinction, which lifted the weight of competition from the mammals. But it was possibly not the only one, because the world changed in multiple ways at the time, for example becoming warmer. The flowering plants also radiated, and their newfound diversity enabled diverse mammalian diets and lifestyles.[31] We may never know whether one or more than one kind of change triggered the success of mammals, but that is beside the point. The point is that their success was triggered by a changing world.

In parallel with the mammals, another group of now extremely successful animals advanced in fits and starts. These animals, however, did not compete with the dinosaurs. They *are* dinosaurs – the last surviving ones.

Among them are tiny but scintillating hummingbirds that whir from flower to tropical flower, sipping nectar with their slender beaks; bar-headed geese that can migrate across the Himalayas in air so rarefied that it would kill other animals; Emperor penguins swaddled in layers of protective fat and down feathers, which enable them to breed in temperatures of minus forty degrees and waddle more than one hundred kilometres across the Antarctic ice; and grey gulls which survive in what may be the most hostile place on earth, the searing heat of Chile's waterless Atacama Desert.

Evolution began to assemble the architecture of birds like these – what biologists call a body plan – piece by piece more than 150 million years ago in the Jurassic period, long before today's nine thousand plus bird species emerged. This was a time when pterosaurs still ruled the sky. Also at that time lived the iconic animal known as *Archaeopteryx*, which amalgamates some bird-like features – feathers, wings – with more reptilian traits like a full set of teeth and three claws protruding from the ends of its feathered arms.

Fig. 4 A fossil of the *Archaeopteryx*

For a long time since its discovery in 1860, *Archaeopteryx* was *the* missing link between birds and reptiles, but recent decades have seen the discovery of multiple other fossils like it. Some are younger than this chimaera; others are even older. Together they rewrote the story of birds and demonstrated that *Archaeopteryx* was only one of multiple extinct evolutionary experiments in birdness.[32] During more than a hundred million years, evolution tinkered with typical bird traits like toothless beaks, eggs, feathers and wings. Forelimbs became longer and more powerful, bones became hollow to reduce weight. They also rotated, twisted and fused to enable a flapping wing. Collarbones merged into a furcula – better known as a wishbone – that absorbs some of the forces of flying. The breastbone grew a keel to attach the powerful flight muscles. Feathers changed from downy filigrees to broad vanes with a sturdy quill. The animals grew smaller, started to walk on two legs, and stopped burying their eggs like most reptiles.

All these changes involved multiple false starts and dead ends, when evolution mixed and matched reptilian and bird-like features in experimental animals that eventually perished like *Archaeopteryx*. One of them was the 130-million-year-old *Mei long*, which is Chinese for sleeping dragon. It still had teeth, but remarkably it roosted like a modern bird, with legs folded underneath its body and beak tucked under one of its wings. Its sleeping position is more than a curiosity, because it hints that *Mei long* was warm-blooded – today's birds sleep like this to preserve heat.[33] Even more bird-like was the 75-million-year-old *Oviraptor*, which sported a toothless beak and brooded its eggs like modern birds.[34]

Some of these evolutionary experiments also created innovative features we find neither in reptiles nor in today's birds. Actually, they were perhaps not features but rather bugs, because they did not survive in modern birds. They might have been a burden rather than a benefit. One of them is embodied in the 120-million-year-old *Microraptor*, which had four instead of two wings, because both its arms and legs were covered with vaned feathers.[35] Four wings might help to create lift, but surely made for awkward walking, because the feathers that covered the *Microraptor*'s legs were so long that they must have dragged on the ground.[36]

The upshot is that some assembly (and experimentation) was required to create the body plan of birds, but that work would be finished millions of years before the other dinosaurs went extinct at the end of the Cretaceous. Alas, the new body plan was not an instant success. Birds did not radiate explosively, at least not right away. They probably were not even very abundant back then, because bird fossils from that time are scarce.[37] (We may of course discover more bird fossils from that time, but even if we do, their numbers would still be dwarfed by those of other species.) Birds did not take off – figuratively speaking – before the other dinosaurs had gone extinct.

And when they did, in what has been called the birds' 'Big Bang', the dinosaur extinction was again not the only cause.[38] That's because the fifteen-kilometre-wide asteroid that careened into the Yucatán peninsula with the force of ten billion Hiroshima bombs did much more than drive the dinosaurs over the edge. It altered the climate worldwide, and helped create a world in which birds could thrive. The debris it ejected covered the skies for years, cooled the atmosphere and dimmed the planet. Forests collapsed worldwide and took a thousand years to recover. Their collapse extinguished the few tree-dwelling bird species that had emerged at the time, but it also created new living space after forests had recovered.[39] This living space would eventually be occupied by those birds, including formerly ground-dwelling ones, that rose once again to the challenge of life in trees.

The point is that no single new aspect of birdness – no one key innovation[40] – triggered the success of birds, because their main features were already in place long before their Big Bang. Instead, a changing environment catapulted the birds from long-lasting obscurity to tremendous success.

The pattern of evolution I describe in grasses, cacti, mammals and birds is so frequent that palaeontologists invented a technical term for it: a *macroevolutionary lag.* Macroevolution is the part of evolution that creates entire branches of new species on the tree of life. (It contrasts with microevolution, which merely alters individuals within one species.) The 'lag' refers to a delay between the origin of a new kind of organism and its eventual success, when an organism either expands in numbers, radiates into many species, or does both.[41]

During a macroevolutionary lag, while an organism scrapes by for millions of years, evolution may change the new organism a little or a lot. It may experiment with the new body plan and refine it, as it did with birds. Or it may repeatedly create and discard similar variations of the same body plan, as it did in tree-living mammals. At the end of the lag period, the organism erupts in new forms, but this eruption is usually not triggered by an innovation. It is triggered by an outside event, a changing environment like the drying of the planet or the extinction of dinosaurs.[42]

Multiple other organisms persevered through similar trials and tribulations. Among them were ants, termites and bees. The first ants, for example, go back 140 million years.[43] However, ants did not begin to branch into today's more than eleven thousand species until forty million years later. And for *another* forty million years, their numbers remained modest. During this time, ant fossils make up a mere one percent of all insect fossils. In contrast, after the ascent of ants thirty million years ago, they account for twenty to forty percent of insect fossils.[44]

We are not certain what triggered their success, but it coincides conspicuously with the rise of one group of plants. This was not just any group, but the single most successful group, the flowering plants, which eventually radiated into almost three hundred thousand species. Flowering plants built the forests where most ants still live today, especially tropical ones, which are both enormous and diverse.

The plant diversity in these forests created countless opportunities for new ways of ant life. A forest with great plant diversity can sustain great insect diversity. Because these insects are food sources for ants, such a forest can also sustain a great number of ant species with unique diets and lifestyles. Some of these lifestyles are truly one-of-a-kind. Among them is 'dairy' farming, which ant evolution arguably discovered millions of years before us. Ants that practise such farming cultivate colonies of sap-sucking aphids and milk them for honeydew. Like any good farmer would, they also protect their aphid herds against predators. Other ant species cultivate enormous fungus gardens that pre-digest leaves for them. These are only two of many myriad ant lifestyles that the ascending flowering plants made possible.

Success was also delayed on smaller, more obscure branches of life's tree. One of them sprouted species of Antarctic fish with a forbidding name: *Notothenioidae*. Amazingly, these fish thrive in sub-freezing temperatures, where growing ice crystals can slice through delicate cells and tissues like knives. They survive thanks to special kinds of proteins that originated more than twenty-two million years ago, and that prevent ice crystals from growing. If such antifreeze proteins are not an important innovation, I don't know what is, yet it would take the nototheni-oids another ten million years before they started radiating. The trigger was once again a changing climate. Before their radiation, the global climate had been several degrees warmer than today, but then, about fourteen million years ago, it started to cool gradually, and created Antarctica as we know it. Ice sheets and glaciers advanced, scoured the ocean floor as they did, and wiped out its former inhabitants, which allowed the newcomers to take over the newly freed real estate.[45]

Even more obscure but no less remarkable is a family of salt water clams known as the *Lucinidae*. Their obscurity is not just figurative but literal, because they burrow under the muddy sea floor. What's remarkable about them is how long they persevered in nearly complete insignificance after they first originated some 420 million years ago in the Silurian period. That was long before the dinosaurs appeared, even before any reptiles or even amphibians showed up. It was a time when the world was a playground for fish, and when plants had barely begun to conquer land.

The *Lucinidae* humbly insisted on surviving for some 350 million years, until they burst into five hundred species some seventy million years ago. To understand why, it's useful to know that these clams partake in an unusual threesome that does not involve sex but mutual feeding with two other species. The first partner is a bacterium that can harvest energy from toxic sulphides in the ocean floor and detoxify them in the process. The bacterium lives in the clams' gills and uses the harvested energy to produce sugars that help both the clam and itself grow.

Enter the second partner, seagrass, which grows expansive underwater meadows that shelter many animals, among them the *Lucinidae*. Seagrasses are themselves remarkable, because their ancestors were land-living flowering plants that recolonised the water, a bit like dolphins, which evolved from landbound mammals. The ancestors of seagrasses began to get their feet wet again some eighty million years ago, and soon formed a peculiar alliance with the Lucinids.

Like a meadow on land, seagrasses protect the ground from erosion and thus protect the clam's habitat. Seagrass roots also release oxygen waste into the mud, which helps the clams respire and grow. In return, the clams and their bacteria detoxify sulphides, which would otherwise slowly poison seagrasses. In other words, the Lucinids helped seagrasses survive. Their help allowed the seagrasses to spread, which enabled the long-delayed success of the Lucinids.[46]

* * *

The further we go back in time, the more fossils have been destroyed by the slow grinding of the continental crust, so fossils can teach us less and less about life's earliest history. However, the teeth of time have left some fossil morsels untouched that are even older than the four hundred million years of the *Lucinidae*. And these fossils reveal perhaps the most remarkable case of delayed success in life's history. It affects not just a single family but an entire kingdom of organisms: the animals.[47]

Molecular clocks tell us that animals originated some 750 million years ago, during a geological epoch known as the Proterozoic.[48] That's Greek for 'former life', and not a bad name, because just about all animals alive then eventually went extinct. And even while they persisted, they evolved few new forms for an unimaginably long two hundred million years, fewer than a snorkeller could discover on a single outing to a coral reef.

Some of these forms resemble today's sponges. Others were elongated tubes permanently attached to the sea floor, and still others resembled fern-like fronds whose undulations must have absorbed nutrients from passing currents. The seafloor itself was covered by dense mats of microbes on which some animals slowly crawled and grazed.[49] However, calling these animals worms would be an insult even to an earthworm, whose body has a sophisticated architecture, with muscles, a nervous system, blood vessels and a gut. To appreciate how simple they were, we have to turn to one of the most mysterious and primitive animals alive today, *Trichoplax adhaerens*.

Trichoplax moves like a single-celled amoeba, even though it consists of a few thousand cells. It is at most a few millimetres long, transparent and so thin – less than a tenth of a millimetre – that it is hard to see with the naked eye. Because of its protean shape, it has neither front nor rear, neither left nor right. Because *Trichoplax* is small it can survive without a gut and blood vessels to transport food. Lacking a nervous system and muscles, it creeps across the ocean floor through the beating of tiny, hair-like *cilia* – the Latin word for eyelashes, even though these lashes propel rather than protect. When *Trichoplax* feeds, it lifts part of its body, a bit like a suction cup that creates a tiny hollow space between itself and the ocean floor – think: outdoor stomach – where it secretes enzymes that digest cells trapped in this space.

Trichoplax is mysterious, because we cannot say much about its evolutionary history. Like extinct Proterozoic bottom feeders, it has no close living relatives and sits alone on a long branch of life's tree. However, the DNA of its genome and its body plan teach us that it is related to sponge-like organisms that emerged close to the origin of animals.[50]

Simple animals like these dominated the Proterozoic. And if a long-lived observer had watched their evolution for millennium after millennium during more than two hundred million years, she would have seen little change among them, certainly much less than in any period afterwards. Based on that dull monotone, she might write the prospects of animals off. If animals subsisted with little more sophistication than slime moulds for all this time, what were the chances they would be destined for greatness?

The chances may have been slim but they were not nil, because eventually animals hit the jackpot. All of a sudden, a bit more than five hundred million years ago, animal diversity increased dramatically in what is called the Cambrian explosion.[51] The Cambrian period in geology is named after the Latin word for Wales, which hosts important rocks from that era. It lasted fifty-six million years, but the major animal groups we know today emerged long before its end, in a small fraction of the time since animals originated. These groups include the arthropods, which comprise organisms as varied as shrimp, spiders, bees and butterflies. They also include molluscs and echinoderms, best known from sea urchins and sea stars. And they include all the vertebrates.[52]

Remarkably, the earliest animals in the Cambrian fauna are nothing like today's animals. They are a veritable freak show of animal diversity, first made popular by palaeontologist Stephen Jay Gould in his book *Wonderful Life*. Among them is a bizarre creature called *Diania cactiformis*, also known as the walking cactus, whose body was elongate like that of a centipede but heavily armoured. It sprouted ten pairs of stilt-like legs covered with prickly spines, as well as a head shaped vaguely like a fluorescent light bulb.

In contrast to *Diania*, *Wiwaxia* crawled over the sea floor with a single slug-like foot. Its helmet-shaped body was covered in armour-plated scales, from which dagger-like spines protruded to deter would-be predators. *Wiwaxia* looked a bit like an ocean-going cross between a hedgehog and a snail.[53]

No less bizarre was *Opabinia*, with multiple pairs of flipper-like flaps equipped with gills. Add to that a bulbous head with five eyes and a long proboscis, a bit like the hose of a vacuum cleaner, but with a single long claw on its end, and you get the picture.[54] When palaeontologist Harry Whittington first presented it at a scientific meeting in the 1970s, people burst out laughing.[55]

Compared to *Opabinia*, which measured only a few centi-metres, *Anomalocaris* – Latin for the 'abnormal shrimp' – was a true giant. At a length exceeding one metre, it was probably one of the top predators at the time. Like *Opabinia*, it swam with multiple pairs of lobes that formed a fanned tail at its end. Its disk-like mouth resembled a pineapple ring formed by multiple overlapping plates that could widen and constrict, like an iris. To the left and right of the mouth, two grotesquely large spiny arms or antennae protruded. Its compound eyes with sixteen thousand individual lenses rivalled the acuity of the best insect eyes we know.[56]

Palaeontologists debate to this day how these and dozens of other bizarre organisms were related to each other and to life as we know it today. The principal problem is that no obvious intermediate fossils exist that could illuminate evolution's path from one to the other. This lack of missing links holds an im-portant lesson: it tells us that these animals arose rapidly, too rapidly for the intermediates to leave traces in the fossil record. In other words, their rapid appearance underscores evolution's creative prowess. And soon after their appearance, they went ex-tinct again, to be replaced by today's fauna. Their replacement makes the same point. Evolution is not at a loss for new life forms when old ones become obsolete.

Which brings us to a key question: what stopped evolution from show-casing its creative prowess much earlier? What de-layed, for some two hundred million years, a revolution that led from near-microscopic brainless bottom grazers to metre-long sharp-eyed predators? Controversy is swirling around this question, and that's not surprising. The Cambrian explo-sion occurred half a billion years ago, and since then most of its traces have been washed away by time. We cannot turn back the clock and observe what held up the Cambrian explosion for so long. However, a good candidate for that show-stopper exists. It is the chemical element oxygen. More precisely, the lack thereof.[57]

Before evolution had come up with photosynthesis at least 2.4 billion years ago, the world's atmosphere was a mix of gases that would have been highly toxic to animals. Most notably, it lacked oxygen. And the oceans would have been just as hostile, because they too lacked oxygen. Even after cyanobacteria and algae had started to exhale this waste product of photosynthesis, oxygen accumulated only slowly. It would take almost two billion years to accumulate in quantities sufficient to sustain large animals.

Oxygen is a source of energy so rich that it is rivalled by only two other chemical elements, chlorine and fluorine, both of which are too unstable to sustain life on earth.[58] A metabolism that is aerobic – running on oxygen – can help release ten times more energy from every morsel of food than a metabolism that is anaerobic. Even today, regions of the globe that are devoid of oxygen, like some deep ocean basins, do not sustain large animal life. They are inhabited mostly by microorganisms.[59] (Abundant oxygen also provides other perks, such as the atmosphere's ozone layer that protects life from damaging radiation.)

Oxygen is such an important fuel that the time needed to fill a planet's atmosphere and oceans with oxygen may govern where in the cosmos complex life can emerge. On earth, this oxygenation took forever – 3.9 billion years, to be precise. That's most of the time since earth's origin 4.5 billion years ago, and almost half of the useful lifespan of the sun's nuclear furnace. Other stars, especially massive ones, can be much shorter-lived, with lifetimes measured in millions instead of billions of years. Such stars may not provide their planets with enough time to become oxygenated, such that planet-hunters may waste their time looking for complex life near them.[60]

The torpid Precambrian fauna is typical of a world with little or no oxygen. In this world, most familiar kinds of animals, such as large, fast-moving predators – even ones much smaller than *Anomalocaris* – cannot obtain enough energy to survive. But eventually the Precambrian ended, and two new kinds of fossils heralded its end, even before the Cambrian explosion began. They were much less dramatic than the Cambrian freak show, but even more illuminating, because they underlined the environment's crucial role in evolutionary success.

The first kind is not even a fossil of an animal body. It is a *trace fossil*. When an animal crawls over the ocean floor, it leaves a trace, and sometimes this trace becomes fossilised. Until the Cambrian, animals left mostly horizontal traces as they grazed on microbial mats. In other words, they did not burrow vertically into the ground. That's how things remained for some two hundred million years, possibly because of a lack of energy, until just before the Cambrian explosion. That's when animals literally began to stir up the muck. Their newly vertical traces prove that they burrowed into the ocean floor,

Fig. 5 A trace fossil of *Fustiglyphus annulatus*

possibly to hide from predators, possibly to sneakily prey on others from below.

What happened then is familiar from gardens tilled by earthworms. By loosening the soil, burrowing animals not only created living space for themselves, they also brought nutrients to the surface that could benefit others. Burrowing created new habitats that could sustain new forms of life.[61] The burrowing lifestyle also reveals that animals were not just passive recipients of oxygen's energetic blessings – they actively participated in creating a new world.

The second herald of the Cambrian explosion was a new group of small animals with shells – the first shelled animals ever. Their fossils appeared in great numbers just before the Cambrian explosion.[62] No larger than a few millimetres, they came in multiple shapes, some resembling the shells of snails and oysters.

Today we can find shells on any walk on a seashore. Their abundance is unremarkable. But at the time they were revolutionary innovations that protected the tender flesh inside. They reveal that predators large enough to eat this flesh had arisen.[63] And they teach us that an arms race between predators and their prey had begun. Within a few million years, this race would eventually create not just bristling defences like that of the 'walking cactus', but also large predators like *Anomalocaris*.

When the Greek philosopher Heraclitus said that strife is the father of all things, he knew neither about evolution nor about the Cambrian explosion. But his words seem clairvoyant to an evolutionary biologist, because strife between predator and prey, made possible by energy-driven activity, was crucial to helping create the large animals of the Cambrian explosion.

Oxygen is a prime candidate for the trigger of this arms race and the Cambrian explosion, but it is not the only one. Some scientists favour changes in temperature, others changes in greater nutrient availability. The identity of this trigger, however, is less important than this: the outside world was crucial to awakening the dormant potential of a new way of life, the animal way of life that had originated two hundred million years earlier.[64]

The Cambrian explosion is highly visible to those who know how to read fossils, but other, more obscure innovations were no less momentous. They go back even further than the Cambrian, to a time from which even fewer fossils have been preserved. One of these innovations has intrigued biologists for decades. It occurred long before the Cambrian explosion but was essential for it. It is the innovation of multicellular organisms – creatures with more than one cell. Their origin poses a profound problem that has fascinated biologists. That's because individual cells in a multicellular organisms must sacrifice their own benefits for the greater good of the whole. Such altruism is hard to understand in a Darwinian world where only self-interest seems to matter for survival.

In a world of single-celled organisms, every cell needs to grow and divide fast enough to outcompete others, or else it is doomed to extinction. Multicellular organisms do not live by the same rules. Multicellularity requires that cells give up their selfish ways, but more than that, it requires that many cells give up life itself. That's because cells have two different roles in most multicellular organisms. One kind of cell stops dividing and specialises to serve others, by performing one of the many jobs needed to keep a body alive: absorbing nutrients in the gut, ferrying oxygen to organs, defending the body against diseases, detoxing blood and so on. This kind of cell dies with the body it serves. The other kind of cell – a reproductive cell like a sperm or egg cell – keeps dividing. Only reproductive cells get passed on from one generation to the next. Only they have a shot at unending life.

What could compel all other cells – trillions of them in a body like ours – to give up life itself? One answer is that multicellularity and self-sacrifice can succeed if the success of one cell – a reproductive cell – becomes the success of all others. And that is the case when all cells in a body share the same DNA, because it's the DNA's evolutionary success that matters above everything. Multicellularity can succeed when multicellular organisms consist of cells with identical DNA.

For this insight, we have to thank the twentieth-century biologist Bill Hamilton who proposed the genetic origins of self-sacrifice in the 1960s.[65] Until then biologists had struggled to explain these origins, seemingly more so than evolution struggled to discover multicellularity. That's because fossils of ancient multicellular organisms – scarce but not absent – first appeared on multiple branches of life's tree that were distant from each other. For example, animals and plants discovered multicellularity independently from each other. Altogether, fossils tell us that multicellularity has evolved independently at least twenty-two different times.[66]

Just as remarkable is another kind of evidence that makes the same point: if the discovery of multicellularity were difficult in evolution, we would never observe it in the laboratory. It would simply require too much time, perhaps many million years. But the opposite is true. Primitive forms of multicellularity can evolve in the right kind of experiment. And they need mere weeks to do so.

One such experiment started from a single-celled algae called *Chlorella vulgaris*, which stubbornly clings to its single-celled way of life when kept by itself in the laboratory. That changes when *Chlorella* gets unwelcome company in the form of another single-celled organism called *Ochromonas vallescia*. This one is a predator that engulfs its prey to swallow it whole. When *Ochromonas* is first let loose on a population of *Chlorella*, it devastates the algal population. But once it has eaten most algae, the predator begins to starve and its numbers decline. That decline gives the algae a chance to bounce back. And when they do, they do not come back as single cells, but as large multicellular colonies that reproduce by budding off daughter colonies from a mother colony. The main benefit of such a colony is size. The large colonies are too big for the predator, which can no longer engulf and eat them. Remarkably, this primitive form of multicellularity emerges in the laboratory within a mere twenty days.[67]

Another striking experiment was performed in 2011 by Will Ratcliffe and collaborators, then at the University of Minnesota. They worked with brewer's yeast, a single-celled fungus used to brew beer and bake bread. In their experiment, they allowed a yeast culture in a liquid nutrient to stand on a benchtop for forty-five minutes without being moved or shaken. During these forty-five minutes, some yeast cells remained floating in their nutrient broth, whereas others settled on the bottom of the container. The researchers then transferred only the settled cells into a new container with fresh nutrient broth. They allowed these survivors to grow and divide for a day, shaking their culture constantly so as to aerate it. After that, they allowed the cells to settle again for forty-five minutes, and repeated the whole cycle – selection, growth, settling – for sixty days.[68]

After this time, yeast cells were no longer single-celled, but had evolved to form multi-cell clusters resembling snowflakes. Because these clusters harbour many cells and are massive, they settle rapidly, increasing their chance to survive into the next cycle of growth and selection. The clusters had also evolved a primitive mode of reproduction. They divided in two by splitting in places where cells had died – possibly sacrificing themselves – which broke the connection between two clusters.[69] When the researchers repeated this experiment multiple times, these cell clusters emerged every single time.

While Precambrian evolution may never have favoured organisms that drop to the bottom of a test tube, the success of such unnatural selection makes an important point. There are multiple routes to multicellularity, and in the right environment – here created by an experimenter – it can rapidly emerge and become successful. Nature does not have to wait long for the right DNA mutations, like those that help cells stick together or sever the ties between them at the right time.

If we know little about the origins of multicellularity, we know even less about the most momentous innovation on this planet, the origin of life itself. However, tantalising hints suggest that it, too, may be easier than we think, and that it, too, was helped along by a changing environment. Exhibit A is that multiple essential building blocks of life can easily form in nature, all by themselves. That first became clear in an iconic 1952 experiment that aimed to recreate the chemical environment of our planet some four billion years ago. In this experiment, chemist Stanley Miller mixed likely ingredients of a primordial soup, including water, methane and ammonia. When he energised the mix with heat and lightning – electric sparks – multiple amino acid building blocks of proteins formed within a week. More recent experiments created multiple other such building blocks, most notably those of RNA, which preceded DNA as the substrate of inheritance in early life.[70] What is more, many of life's building blocks can form in a place much more hostile than early earth: outer space. We learned that in 1969, when a now famous meteorite, as old as the earth itself, descended from space and careened into a barn in the Australian hamlet of Murchison. Chemical analysis showed that it ferried multiple building blocks of life.[71]

Although we have not yet succeeded in creating life from these building blocks, laboratory experiments are making steady progress. For example, they recently showed how early life may have solved a problem that is as crucial for life's origin as for life's subsequent evolution. This is the problem of reproduction.

Reproduction requires that one cell can divide in two. It is essential for Darwinian evolution, because without reproduction, natural selection cannot work. Reproduction is a sophisticated process in today's cells, controlled by dozens of proteins. That's why its origins seem hard to explain. However, recent experiments show that these origins may be surprisingly simple.

Cells are essentially bags of molecules and what holds these molecules together – the bag itself – is a lipid membrane, a thin layer of elongate molecules aligned with one another and forming a two-dimensional sheet. Such a membrane, it turns out, can form in a test tube, all by itself, if the right kinds of molecules are present. These molecules include fatty acids, whose precursors also happen to be found on meteorites.[72] What is more, fatty acids can form not just membranes but vesicles, closed spheres that resemble cells. These vesicles can grow by adding more fatty acids or by merging with other vesicles. And in the right conditions, for example when they grow rapidly, or when the water around them evaporates, they lose their spherical shape and turn into elongate tubes. Highly unstable tubes, I should say, because the slightest agitation will cause them to fragment into multiple vesicles, subdividing the tube's molecules among these vesicles. In other words, bags of molecules are not only able to assemble, but to grow and divide all by themselves. They do not need the sophisticated controls of modern cells, even though these controls may ensure a precise timing of growth and reproduction.[73]

Unfortunately, molecules like fatty acids would have found it hard to persist – let alone self-assemble – shortly after the violent birth of our planet 4.5 billion years ago. That's when a cataclysmic crash of two celestial bodies created both the earth and the moon, releasing enough energy to boil rocks. The result was an early atmosphere filled with rock vapour, which eventually cooled enough to liquefy and rain down into magma oceans. Although the heat of this magma eventually radiated into space and created the solid crust we stand on, the atmosphere, which then consisted mostly of carbon dioxide, trapped enough solar heat to keep the surface temperature above two hundred degrees Celsius. What is more, for at least a few hundred million years a hail of meteorites known as the Late Heavy Bombardment rained down on earth and the moon. You can still admire its scars as lunar craters. The diameter of some of these rocks exceeded three hundred kilometres, twenty-five times larger than the one that wiped out the dinosaurs, each liberating enough energy to sterilise the entire planet multiple times over.[74]

This inferno subsided sometime between 4.2 and 3.8 billion years ago. That period of time is fascinating, because it's also when the first signs of life appear. These signs are no fossils of living bodies. They are not even trace fossils. They are chemofossils – chemical signatures of life. One of them comes from carbon isotopes, stable forms of carbon, such as ^{13}C (pronounced carbon thirteen), which is heavier than the lighter ^{12}C most carbon consists of. Enzymes prefer to use the lighter isotope when they build cells and organisms. This means that organisms are richer in ^{12}C than the inanimate world around them. And some of the most ancient rocks bear precisely this kind of signature. They are enriched in ^{12}C, which goes to show that some of their carbon comes from long-perished living beings. Their age places the origin of life sometime between 3.7 and 4.2 billion years ago. In other words, life arose almost as soon as it could. It arose when the environment became placid enough for molecules to self-assemble into structures like the growing and dividing vesicles that also form by themselves in the laboratory.[75]

In sum, multiple events in life history immediately followed the creation of the right environment, including the radiation of lupins that followed the uplift of the Andes, and that of cichlids, which tracked the creation of East African lakes. The earliest chemofossils tell us that the origin of life was only the earliest and most seminal of these events. Add to that the multiple independent origins of most innovations, such as multicellularity, mammalian ways of life or C_4 photosynthesis, and it is hard to escape the conclusion that evolution is rarely at a loss for new solutions to life's problems. To the contrary, life often innovates before the time is ripe, before the environment is ready for an innovation's success. That's what we learn from the early history of organisms as different as grasses, mammals and ants, which innovated in obscurity or merely persisted for millions of years before their time came. Next, we will zoom in from whole organisms to cells and their molecules. They will help us understand some of the causes for evolution's talent to create the new and useful before its time.

3

Molecular motorways

THE BACTERIUM *SPHINGOMONAS CHLOROPHENOLICUM* HAS an unusual superpower, so striking that it is immortalised in the bug's name. This name alludes to pentachlorophenol, a dangerous chemical that was first sold in the 1930s as a pesticide. However, it poisons more than just pests. It also damages the kidneys, blood and DNA of animals, including humans. But its toxicity does not faze *S. chlorophenolicum*. The bacterium can survive on pentachlorophenol. And it can do more than that. *S. chlorophenolicum* can extract all the energy it needs from the toxin's atomic bonds. As if that were not enough, it can also build the carbon backbone of all molecules inside its cell – billions of them – from the carbon atoms in pentachlorophenol.[1] *S. chlorophenolicum* has evolved the remarkable ability to turn a toxin into food.

We already encountered rapid evolutionary changes like the loss of stickleback armour and the radiation of lupins, but this innovation beats them all. It appeared in a blink of evolutionary time, within the century since humans first discovered pentachlorophenol. Its appearance was triggered by the environmental challenge of a man-made chemical. And it's especially interesting because we know why evolution could rise to this challenge so rapidly.

Like most bacteria, *S. chlorophenolicum* has a notable talent, the ability to incorporate DNA from other cells into its genome.

The process is called *horizontal gene transfer* and it can take various forms.[2] For example, some bacteria can ingest the DNA of other cells that die and rupture. In this molecular version of necrophagia the ingested DNA will often be chopped into little pieces for food, a job that falls to specialised protein enzymes. These and other enzymes protect a cell's genome by cutting up foreign DNA and repairing its own DNA. But sometimes these enzymes make mistakes. They inadvertently insert the foreign DNA into the bacterium's own genome. And if that DNA includes one or more genes, a bacterium has acquired the ability to make new proteins, each with a unique skill that can come in handy at the right time.

Another means to get new DNA is a form of sex that is as unusual to us as it is common in bacteria. It begins when one bacterium builds a long protein tube called a sex pilus that protrudes from its cell wall. When that pilus latches onto another bacterial cell, it becomes a conduit to transfer DNA to the other cell, which can then paste the new DNA text into its genome.

That's not so different from animal sex, you might think, where a penis acts like a sex pilus in helping to transfer DNA between organisms. But this similarity is superficial. The differences are profound. Bacteria neither transfer DNA to reproduce, nor do they always transfer an entire genome, as when a human sperm fertilises an egg cell. Instead, they often transfer short DNA snippets with just a few genes.

Even more importantly, horizontal gene transfer helps bacteria exchange DNA with a promiscuity unrivalled by people or other animals. That's because bacteria can take in genes not just from their own species but from many others. They can acquire DNA as different from their own as human DNA is from that of chimps, birds, crocodiles, salamanders, fish or plants. And they do so relentlessly. The trillions of microbes underneath our feet, on our bodies, and on any surface around us are constantly transferring DNA, shuffling and reshuffling their genomes, creating ever-new combinations of genes. Over time, through multiple transfers, this DNA can quickly add up to staggering amounts. In bacteria like *Escherichia coli*, some strains – variants of the same species – can harbour multiple stretches of foreign DNA, totalling hundreds of thousands of letters, each one of them transferred from elsewhere, each one present in one strain but absent in others.[3]

Gene transfer is important, because it allows a bacterium to draw on diverse tricks other bacteria have evolved to meet life's challenges. It is a form of pilfering knowledge, chemical knowledge encoded in genes and embodied in proteins. But it can achieve more than that. By combining a cell's own embodied knowledge with that gleaned from other organisms, it can create new skills, new chemical means of survival. The technical term is *recombination*. By recombining the DNA of two or more organisms, horizontal gene transfer creates new embodied chemical knowledge. This kind of recombination helped *S. chlorophenolicum* survive on pentachlorophenol. Here is how.

To feed on pentachlorophenol *S. chlorophenolicum* first needs to transform this toxin into a less toxic molecule that is more like a conventional nutrient. To this end, the bacterium uses specialised protein enzymes, each encoded in a gene, each consisting of a unique chain of amino acids, each capable of catalysing a single chemical reaction.[4] But a single chemical reaction does not suffice for this transformation. A whole sequence of such reactions is necessary, what biochemists call a *metabolic pathway*.

Each step of this metabolic pathway has its own enzyme and gene. While some of the genes come from the bacterium's own genome, others have been transferred from different organisms. In other words, horizontal gene transfer has helped the bacterium combine its own chemical reactions with those from another organism to create the new pathway. The bacterium's innovation does not involve new genes, but a new combination of old genes. It is this new combination – recombination – that creates the new pathway which can transform pentachlorophenol into a nutrient.[5]

Horizontal gene transfer incessantly creates new combinations of genes, proteins and chemical reactions in microbes all around us. That's why the microbial world is brimming with invisible innovators like *S. chlorophenolicum*. They can also parry many other chemical blows that humans have inflicted on the world. One such blow is Aroclor, a potent cocktail of toxic molecules sold until the 1970s by Monsanto for electrical insulators and industrial fluids. It contains polychlorinated biphenyls that can mimic hormones, disrupt our immune system and cause tumours.[6] When released in the water, these molecules can also exert horrific effects on fish, which spurt blood, shed skin and turn belly-up within seconds after having been exposed.[7] Because these chemicals are resistant to heat, acids and other efforts to destroy them, the hundred thousand tons or so that are sloshing around in the biosphere would not disappear any time soon, were it not for innovative bacteria that can feed on them.[8] Such bacteria indeed exist, not just one species, but several of them.[9]

Evolution in action can also help neutralise chlorobenzene, a chemical solvent.[10] And it helps destroy DDT, the pesticide infamous because it threatened to wipe out America's national bird, the bald eagle, by thinning its eggshells until they cracked. Fortunately, DDT was banned in America in the 1970s.[11] However, such bans might achieve little unless some microbes can mine DDT, chlorobenzene and other man-made chemicals for energy and building materials, purging them – albeit slowly – from the environment.

None of these evolutionary success stories are widely known, because none of them are inscribed in the bodies of charismatic organisms, like the towering spire of a redwood tree, the quiet menace of a lurking tiger, or the slender grace of a pink flamingo. That's a shame, because most of evolution's hard labour takes place in this invisible world of chemistry. That's where evolution hones the thousands of molecules that keep the cells of bacteria and humans alive. That's also where evolution creates countless biochemical pathways that synthesise new molecules or take apart others to detoxify them. Such chemical creations are not a peculiarity of toxin-munching bacteria. They started with the origin of life, continue to this day, and permeate the biosphere, numerous beyond words. They include complex hydrocarbons that help ants living in giant colonies tell friend from foe by touching each other's bodies.[12] They include pigments that help flowers attract pollinators and that delight us in multi-coloured meadows. They include sex pheromones whose faint trails are irresistible to male moths.[13] They include devious chemical weapons like neurotransmitters that are synthesised and released by plants to confuse the brains of hungry insects and to avoid getting eaten.[14] And they include many molecules we cherish, like the caffeine that boosts our energy and the morphine that relieves our pain.

The further we go back in evolutionary time and the closer we approach life's origin, the more important such chemical innovations become. In life's earliest days, they established life's chemical foundation. This foundation is energy metabolism, a complex network of chemical reactions that connects multiple metabolic pathways to harvest energy from the environment. One of its earliest chemical innovations was the novel chemistry of photosynthesis discovered by cyanobacteria more than two billion years ago. It enabled these bacteria to harvest energy from sunlight, and it liberated the very oxygen that would slowly fill the atmosphere. Another early innovation was the ability to use this oxygen in the process we call respiration, which arose shortly thereafter. As we have already learned, the complex, multicellular life we know would not exist without it.[15]

Between respiration and today's *S. chlorophenolicum* stand countless other innovations in energy metabolism. Each of them empowered life to use a new kind of nutrient. Some of these nutrients are well-known energy-rich molecules, such as the sugar fructose. Others are more arcane, with names like alloxan, uridine or inosine. These innovations slowly accumulated over many million years, and helped create today's species and their metabolisms. That's metabolism in the plural, because each species has a unique evolutionary history that can come with unique metabolic innovations. Many of these species are marvels of metabolic prowess and flexibility.

This prowess is on display in the bacterium *E. coli*. Like *S. chlorophenolicum* and many other bacteria, *E. coli* can use a single kind of nutrient molecule like the sugar glucose to meet all its energy needs. From this kind of nutrient, it can also extract all the carbon atoms to build the molecules that it requires to survive and grow.[16] These essential molecules include the twenty amino acid building blocks of proteins, the four building blocks of its DNA, the lipids of its cell membrane and multiple others, a total of some sixty different kinds of essential molecules, each of them occurring in millions of copies inside each *E. coli* cell.

E. coli's metabolic prowess stands out even more when compared to our own. Our bodies fail to synthesise nine so-called essential amino acids. We need to get them from our diet. If we don't, we die. In addition, we need multiple other vitamins, organic molecules that we cannot assemble either. *E. coli* can build all of them. We think of bacteria as simple organisms, but their metabolism is more powerful than ours – a lot more. Compared to them, we are metabolic cripples.[17]

Just as important, *E. coli* can assemble all essential molecules from not just one nutrient, but from more than seventy *different* kinds of nutrient molecules – sugars like glucose, acids like acetate and alcohols like ethanol.[18] And each of these seventy foods can also provide all of *E. coli's* energy. This bacterium is like an engine that cannot just run on wood, oil, gas or coal, but also on anything from milk to tequila to toilet cleaner. However, to call it a mere engine is to belittle its powers. Human engines just create motion, but this one also manufactures its own parts and replicates itself, all from the energy and carbon contained in any one of seventy different raw materials.

This astounding flexibility has a price: high complexity. The network of metabolic reactions in *E. coli* comprises almost fourteen hundred chemical reactions, each catalysed by a protein enzyme, each encoded in *E. coli's* genome.[19]

Some of these reactions form pathways whose steps transform, like the pentachlorophenol pathway, a molecule into an easily digestible one. Think of these pathways as the alleyways or backroads of a city's road network. They merge into wider streets, sequences of reactions modifying common digestible molecules, cleaving them, combining them or rearranging their atoms. These streets themselves merge with avenues that carry even more molecular traffic. Eventually, most of this traffic ends up in one of a few multi-lane boulevards where the most common chemical reactions take place.

The similarity to a road network does not end there. Like roads, individual pathways can also bifurcate and branch into two sequences of reactions. Larger roads can sprout smaller ones, each leading to an essential molecule that is assembled step by step from smaller parts. And individual pathways can merge in roundabouts from which multiple other pathways emerge.

Metabolism is as complex as the road network of a major city, but not as regular and grid-like as that of Manhattan. Instead, its roadmap resembles that of old cities like Rome or London. As such cities grow during years, decades and centuries, many new roads are added, while others are widened, shortened or obliterated. Roundabouts, speed bumps and one-way streets are created to regulate traffic flow. And this construction work never seems to stop – as soon as one construction site closes, three others seem to open. The same holds for life's metabolic road network. It too has been under construction ever since life began, although without the help of bulldozers, graders and pavers.

The tools of evolution to create new enzymes, pathways and crossroads are DNA mutations and horizontal gene transfer. They help build metabolic pathways that break down new foods, create new and useful molecules, or build such molecules in new ways and from different parts. When a pathway becomes obsolete, perhaps because its food source has permanently run out, the pathway quickly disappears again through DNA mutations that inactivate its genes or erase them from the genome altogether.

Every single cell on this planet, bacterial or human, harbours a self-remodelling metabolic network like that of *E. coli*. Some of these networks are not quite as complex, others even more so, but all of them are completely invisible to our most powerful microscopes. That's why twentieth-century science had to map their reactions and pathways through laborious biochemical experiments. For *E. coli* alone, it took more than half a century and thousands of biochemists to create a nearly complete map. These scientists also mapped countless metabolic pathways and reactions of other organisms. They amassed an enormous amount of metabolic knowledge that eventually triggered an information revolution. It allowed metabolic biochemistry to enter the era of Big Data.

Computers are revolutionising biology in the twenty-first century much like microscopes did in the seventeenth. They are giving us access to an invisible world that is as plain to them as a road map is to us. They are especially valuable for metabolic analysis, where they process metabolic information assembled by thousands of biochemists, and combine it with the output of machines – DNA sequencers – that can read the millions of DNA letters in an entire genome overnight. In doing so, they can automatically assemble nearly complete metabolic roadmaps of previously unstudied organisms. And they can do more. They can tell us what a metabolic network is capable of, what food sources it can use, what molecules it can synthesise, and how much molecular traffic it can support.

It's not hard to understand the principle behind the algorithms, the recipes, for such computations. To begin with, an algorithm needs information about an organism's chemical environment, including all the available food molecules. It also needs information about the organism's enzymes and the chemical reactions these enzymes catalyse. With this information, it can tell us which new molecules the enzymes can produce from the available food. Some of these new molecules can then react with one another if the organism has the right enzymes. These reactions create further molecules, which can then react with one another, and so forth, creating wave after wave of new molecules. A computer can work out all the molecules in each of these waves. And it can work out whether all of life's essential building blocks are among them. If so, we know that the organism can survive on the available food.

That's the idea. Its implementation is more complicated, but it can also give us more information. For example, it can tell us how much of each essential molecule an organism can build, and how fast the organism can build it. In other words, it can tell us how fast an organism can grow and divide.

By comparison, our experimental technologies lag far behind. While they can measure the concentration of a few hundred molecules at a time, they get nowhere close to mapping all molecular traffic in a cell. But experiments can be helpful in other ways. They can help us find out how well our computer algorithms perform. One helpful kind of experiment uses genetic engineering to mutate and inactivate a specific enzyme. The knocked-out enzyme can no longer perform its job – to catalyse a specific chemical reaction. Such an experiment effectively creates a metabolic roadblock. If the blocked road carries little traffic and can be easily detoured, metabolism will be undisturbed and cells will continue to grow rapidly. If the roadblock backs up traffic elsewhere in the network, it will slow a cell's growth down. And if the roadblock cuts a trunk road, it will kill the cell. After installing such a roadblock, one can measure whether a cell grows more slowly or not at all.

Repeat this experiment with many different enzymes, knock each one out, measure its effect on growth, compare the effect to a computer's prediction, and you find that the computer does a great job. Its predictions agree with experiments for ninety percent of engineered roadblocks, not a small feat considering the complexity of metabolism.[20] Where its predictions do not agree, the fault often lies with our incomplete knowledge of the metabolic roadmap.

Researchers around the world and in my Zürich laboratory routinely use such computations without giving them much thought. We rarely consider that they are among the greatest achievements of biology since Charles Darwin's *Origin of Species*. A central concept in Darwin's theory is fitness, and for an organism to be fit – adapted to an environment – it is essential that the organism is viable in that environment. That is, it must be able to survive. Darwin and many others after him had to measure viability in often difficult experiments. With our new computational instruments, however, we can begin to compute it, at least for microbes. What is more, we can compute it in split-seconds, and from first chemical principles, from the chemical reactions in a metabolic network.

And we can do much more. We can ask whether a cell's growth and fitness can be improved by tweaking individual pathways, widening them or moving intersections, thus allowing traffic to flow better and reduce congestion. We can also find out which metabolic innovations – new reactions and pathways – would allow a metabolism to survive on new foods or produce new and useful molecules.

I already mentioned why this is possible: biochemists have not just studied *E. coli*'s metabolism but that of multiple other organisms, most of them equipped with unique enzymes found neither in *E. coli* nor anywhere else. As a result of their labour, we know some ten thousand metabolic reactions, each of them taking place in some organism somewhere in the biosphere, each of them encoded in genes that wait to be recombined with others to create metabolic innovations. All these reactions are held in public databases like the Kyoto Encyclopaedia of Genes and Genomes, ready to be downloaded by anybody and everybody.[21] Our computers can combine and recombine these reactions just like nature would through horizontal gene transfer, exploring different pathways until they find one that allows a cell to digest new foods or build new molecules. For the first time in history, innovation too is becoming computable.

And we can do something more extreme. We can task our computers to design a metabolism from scratch, specifying only the nutrients it can use and the molecules it must produce. The task is to build a chemical network creating these products from the nutrients we specify. The network must use only chemical reactions and pathways that exist in the biosphere. In 2009, my research group first began to develop a computer algorithm that could solve this task. This algorithm modifies the known pathways of a metabolism, either adding known reactions from the biosphere, or deleting reactions and rerouting the metabolic road network around these road blocks. It works a bit like horizontal gene transfer, but more radically, by changing a metabolism's pathways in thousands of small steps. In this way, the algorithm explores myriad new ways to digest food, produce energy and assemble molecules, until it has created a metabolism that does the job we need it to do.[22]

We first applied this algorithm to create a metabolism with capabilities like that of E. coli, digesting similar foods and producing the same molecules. It was a trial run, and I had no great expectations, thinking that it would probably design a metabolism very similar to E. coli. But boy, was I in for a surprise. The metabolism it constructed did not take after E. coli. Most of its reactions and pathways differed, even though its chemical capabilities were similar. What is more, when we ran the algorithm again, it came up with yet another metabolism that resembled neither the first one nor E. coli. We ran it a third, fourth and fifth time, and then over and over again, hundreds of times. Each time it produced a metabolism unlike any of the others, sharing only a minority of reactions with any one of them.

The lesson was clear: many different metabolisms are endowed with the same capabilities. Naturally, we wondered how many. But when we tried to compute their number, we failed, and we quickly discovered why. The number is so large that evolution could not produce them all, even if it continued for another four billion years. It is greater than the number of stars in our galaxy, greater even than the number of hydrogen atoms in the universe. It is astronomical beyond comprehension.[23]

This means that metabolically flexible species like E. coli float on the surface of a deeper truth: life's chemistry itself is flexible, and that flexibility enables flexible metabolisms. It endows life with different ways to digest foods, extract their energy and construct new molecules, so many different ways that their total number defies imagination. And this flexibility is more than just a computer's prediction. Researchers who study metabolism in the wild keep discovering new metabolic pathways, especially from bacteria. Life's ingenuity in digesting and assembling the same molecules seems unending.[24]

My previous book *Arrival of the Fittest* explains the profound consequences of such flexibility for evolution. Here, I mention it because it led me to a new and completely unexpected source of metabolic innovation. I discovered it in 2012 with my PhD student Aditya Barve, when I asked him to create a different kind of metabolism designed by computer. We wanted a metabolism that was viable not on dozens of different nutrients like *E. coli*, but on something much simpler. We wanted a metabolism that was viable only on a single nutrient. This nutrient and no other should be able to serve as the only source of carbon and energy.

Our algorithm gamely delivered such a metabolism, but when we checked its handiwork, something did not seem quite right. To be sure, the metabolism was indeed viable on glucose. However, it was also viable on five other nutrients, among them common sugars like maltose, fructose and lactose. Metabolic chemistry gave us more than we had asked for.[25]

Suspecting a fluke, we ran the algorithm again, and it found another metabolism viable on glucose. That one was also viable on additional nutrients, except that these molecules were different from the first ones. We ran the algorithm again and again, hundreds of times, each time getting a metabolism viable on glucose but also on additional nutrients, between one and twenty of them. All in all, their numbers added up to forty-five new kinds of foods. Each one could sustain life, and each one emerged in a metabolism that we required to survive only on one kind of food.[26]

We did not yet understand why, but other researchers had found something similar in laboratory experiments. They had wanted to create bacteria that produce better biofuels, by letting the bacteria grow for many generations on the sugar xylose. This sugar occurs in raw materials for biofuels, but many bacteria grow poorly on xylose, and produce little biofuel from it. The idea was to force the bacteria to grow on xylose for a long time, so that they would evolve to use it more efficiently. The experiments succeeded. The bacteria became proficient at extracting energy and carbon from xylose. But unexpectedly, the evolved bacteria had also become viable on another sugar called arabinose, even though – and that was the real surprise – the environment did not contain this sugar.[27] They had evolved an innovative skill, even though that skill was useless where it had evolved. That skill only becomes valuable if arabinose occurs in the environment. And it becomes life-saving whenever arabinose is the *only* food source in the environment.

Metabolism provides innovations for free, because a metabolism evolved to use one food can also use others. But why? We were puzzled by the reasons for some time, until we found a forehead-slapping simple answer. The traffic analogy can explain it. If you build a house in undeveloped land outside a city, you may have to build a road to connect your property to the city's street network. That road may pass close to the properties of some of your neighbours, and if they have not built a road yet, they will benefit from your road, because it also connects them to the city.

Just like your road passes your neighbours before merging with a city street, a new biochemical pathway transforms food into a series of other molecules before merging with the rest of metabolism. And when that food runs out, each molecule along the path can substitute for it, just like each of your neighbours can use the road you built to connect to the city.

For the same reason, *different* designer metabolisms that perform the same task can give us *different* kinds of innovations for free: the road connecting your home to the city can be built in different ways, each of them passing different neighbours and connecting them to the city.[28]

We also discovered other nuggets when sifting through metabolic data. Scientists often bemoan the complex byzantine architecture of cells, organs and bodies, because it makes our research so much harder. However, we noticed that their complexity is great for metabolic innovation. We learned this from designer metabolisms that perform identical jobs but with a different number of chemical reactions, either more than *E. coli* or fewer – the equivalent of a more or less complex road network. The greater the number of reactions became, the greater the number of free innovations.[29] It's not hard to understand why. The more reactions a metabolism has, the more pathways these reactions form, and the more connected the metabolic road network becomes. Just like the number of homes connected to a city increases with the number of roads, so can the number of life-sustaining molecules increase with the number of metabolic pathways.[30]

The source of all this free innovation is the same: the organisation of life's chemistry into branching, merging and interconnecting pathways, where a few new connections can create many more new abilities, just like a road network that serves not only the people it was built for, but many others.

These benefits of a networked metabolism extend far beyond individual species. They also apply to an entire community of microbes. Microbiologists call such a community a *consortium*. The name itself is revealing. Just as a consortium of people or companies collaborates to accomplish a task, so does a consortium of microbes. In the microbial world, the task is usually to squeeze a maximum amount of resources from the environment in order to survive and grow.[31] Microbial consortia share that goal with scientists who want to clean up the mess that some of our chemical industries have made of the world.

We have already learned about individual bacteria that can clean up toxins such as pentachlorophenol, but entire consortia can do much better. In such a consortium, the enzymes of one species excel at a few of the chemical reactions that neutralise a toxin. These enzymes handle the reactions very efficiently, but get the job only partly done. Another species picks up where the first left off, takes a few further chemical steps, but eventually passes the baton to a third and so on. When the final goal is reached, the toxin has yielded all useful atomic materials, and in the process, has become harmless. Consortia are especially useful whenever toxic waste consists of multiple chemicals, such as when crude oil spills from a tanker or a drilling platform.[32]

Consortia are not just practically useful; they can also help us better understand evolution. For example, they open another window into the evolution of multicellular life. I already mentioned one advantage of multicellularity: a large body, too large for would-be predators to attack and eat. But it is by far not the only advantage. Another is the ability to divvy up the many tasks of surviving among different cells, where each cell is specialised for one task that it performs better than others. Such specialisation and labour sharing is a hallmark of sophisticated animal and plant bodies. However, they were not the first to uncover it. Long before them and independently from them, bacteria stumbled upon simple forms of labour-sharing multicellularity, and they did so multiple times.[33] Among these innovative bacteria are the cyanobacteria. They are best known for their discovery of how to harvest energy from sunlight, and produce oxygen in the process. This discovery was revolutionary, but it also created a massive new problem. The problem revolves around nitrogen, and one of its solutions involves multicellularity.

Life needs nitrogen just like it needs carbon, but nitrogen is much scarcer, and this scarcity puts a brake on life's prosperity. That's why crops planted year after year in the same field eventually leach all nitrogen from the soil, become stunted, grow sickly leaves and produce few seeds or fruits, unless the nitrogen is replenished.

The one place where nitrogen is abundant is our atmosphere, which consists of nearly eighty percent nitrogen. Unfortunately, this form of nitrogen is chemically so stable that it is basically useless. Only in the twentieth century did humans succeed in converting it into more useful ammonium, through a chemical procedure called the Haber-Bosch process. This process helped create nitrogen fertilisers that revolutionised agriculture. However, it requires high pressures, extreme temperatures – above 400 degrees Celsius – and a chemical environment hostile to life. It could emerge only from human ingenuity, not from evolution.

Cyanobacteria found a greener solution, but one that comes with its own trade-off. It is an enzyme called nitrogenase, which creates useful ammonia molecules directly from nitrogen gas. The catch: nitrogenase is extremely sensitive to oxygen, such that the tiniest amounts of oxygen will poison and incapacitate it. That's not a small problem, considering that oxygen is the primary waste product of photosynthesis.

Cyanobacteria have a tough choice: either harvest light energy and starve for nitrogen, or starve for light energy and harvest nitrogen. The problem is a real head-scratcher, until you hear about the ingenious workaround some cyanobacteria found. They build slender, blue-green filaments of multiple cells, and most of these cells use light to create their energy and carbohydrates. However, a few specialise into creating ammonia fertiliser from nitrogen gas. To shelter their delicate nitrogenase, they no longer produce oxygen. More than that, they even encase themselves in a nearly airtight cell wall to avoid oxygen.[34]

These organisms not only display the simplest kind of multicellularity, with only two types of cells, but are also a textbook example of labour-sharing: the fertiliser-producing cells feed ammonia to the photosynthesising cells. In return, they receive energy-loaded carbohydrates from them.

Symbioses like these require millennia of evolution to get going. Or do they? One scientist who has been wondering is Eric Libby. A talkative, effervescent Houstonian with a booming laugh, Eric is himself an exemplar of life's extreme adaptability. That's because his career path took him from muggy Houston to the frigid town of Umeå in northern Sweden, where his research thrives against the odds that a subtropical creature faces in a subarctic environment.

I first met Eric at one of his academic way stations, the Santa Fe Institute for the Study of Complex Systems, where I learned about his deep interest in the origin of multicellular life. Every year this research institute in New Mexico hosts hundreds of scientists from many different disciplines, who stay anywhere between a few days to a few years. On my own annual visits during more than a quarter century, I have encountered many of these scientists and exchanged notes with them. Their knowledge combined with my own – recombination – sparked multiple new ideas that have helped me write books like this one. Some of these reverberated in my mind for days or weeks, and one of them led to a collaboration with Eric.

Eric wondered how hard it is to mould organisms into symbionts like the cyanobacteria we heard about, or like the microbes in an oil-guzzling consortium. In fact, he had a radical thought. What if no evolution is needed, what if many species were primed for symbiosis even before their first encounter?

Eric also realised the main problem with this idea: we cannot test it by experiment. Such an experiment would choose two species and a laboratory-created environment where neither of the species can survive on its own. It would cast both species into the environment, and ask whether they can survive together. The experiment itself is not hard. What's hard – actually, impossible – is being sure that the chosen species had never lived together. How do we know that they have not coexisted somewhere on this planet, sometime in their million-year evolutionary history, and that they evolved the ability to collaborate there and then, taking this ability with them to the laboratory?

We don't, and we can't, and that's the problem. We know mostly about present life, and we cannot exclude with certainty that a species' long evolutionary history influences what we observe. But we have an alternative: we can study metabolisms without a history at all. This is where our designer metabolisms come in. These chemical reaction networks can synthesise life's molecules from common nutrients, but they do not carry millions of years of historical baggage. Instead, they are assembled from scratch using only our accumulated knowledge about life's green chemistry.

To answer Eric's question, we created thousands of different designer metabolisms, viable on some sources of carbon and energy but not on others. Many of them produced waste products like the oxygen in photosynthesis, or like the half-digested toxins left over by one member of a consortium. We then chose two of them at random, paired them in our computers to create a simple consortium of two metabolisms, and allowed them to do what nature does, making the waste of one metabolism available to the other. We then computed whether the consortium could live where its members could not, whether it could manufacture all essential molecules where its members failed to do so. And we created not just one such consortium, but many thousands, each of them matched randomly, each of them challenged to survive in new environments.

And they quickly answered Eric's question.[35] The typical consortium could indeed go where its members could not have gone before. It could survive in novel environments, and not just in one, but in multiple environments where each member would have died on its own. And most important, this new and joint skill emerged spontaneously without being shaped by evolution.

When we explored the most innovative consortia more closely, we also discovered another link to metabolic complexity: the greater the complexity of a consortium's metabolism – the greater its number of reactions – the greater the number of new foods on which the consortium could survive. In hindsight, the reason is obvious: a metabolism with more chemical reactions has more opportunities to turn metabolic waste into useful molecules.

Life's chemistry is so flexible that waste molecules of one cell can frequently help another cell survive. Molecules like these can join the chemical skills of two cells and raise their abilities to a new level. It's as if two craftsmen toiling away in their workshops, each mastering their own skills, each creating their own product, could not easily make a living. But if they pool their skills they can create something new that allows them to survive. One might produce steel blades, the other might produce screws, and together they could build scissors in which a pivot connects two blades. One might produce gravel, the other might produce lime, and together they could produce concrete. This kind of collaborative power is built into the chemical foundations of life.

Consortia benefit from the same flexible chemistry that helps individual species innovate. Whether organisms act alone or together, life provides free innovations to them, spontaneously creating the new and useful, simply as a result of metabolism's networked chemistry. In other words, metabolism harbours an enormous latent innovation potential, spouting innovations before they are needed. Most of these innovations are sleeping beauties. They remain dormant until the right environment arises, an environment with novel chemicals where they become important and perhaps even essential for survival. That's also when natural selection kicks in, improving and remodelling new metabolic roads until they can sustain maximum traffic. As we shall see next, underneath the surface of this network, on the level of individual proteins, another innovative force operates, just as invisible and at least as powerful.

4

Good vibrations

WHEN AN ARMY HELICOPTER FLEW over remote southern Venezuelan jungle in 2008, its crew sighted an isolated Yanomami village that would lead to an astonishing medical discovery.[1] The Yanomami are an Amerindian tribe of some thirty-five thousand people inhabiting more than two hundred villages in the jungles of northern Brazil and southern Venezuela. They are among the dwindling number of indigenous people whose way of life – horticulture, hunting animals, gathering fruit – has not been obliterated by contact with our civilisation. Such contact is often disastrous, for example when gold-diggers enter Yanomami territory and bring deadly diseases like measles and malaria.[2] The village discovered in 2008 had been spared such calamities, because it was uncontacted.

Contact with our civilisation may be dangerous, but it can also improve Yanomami lives. According to one Brazilian government estimate, more than seventy percent of Yanomami deaths could be prevented through basic health care and vaccination.[3] That's why regional governments send medical missions to Yanomami villages. When such a medical mission first entered the village in 2009, scientists sampled stool and skin from the villagers. This was part of an effort to understand how the communities of bacteria inhabiting human bodies affect human health. That's an important question, given that these bacteria

– our microbiome – outnumber our own cells.[4] Many of these
bacteria are useful. They help us digest food, ward off disease,
and keep our weight in check. But a few of them can make us
sick. What is more, some of these bad bugs have an especially
deadly talent, the ability to disarm the antibiotics that doctors
use to keep them under control.

Modern antibiotics started out as miracle drugs in a world
where a child with a minor scrape might die from sepsis, where
countless women succumbed to infections after childbirth, and
where more soldiers perished from disease than from enemy
fire. But a mere eight years after penicillin, the first modern antibiotic, had become available in 1945, bacteria resistant to it had
already spread worldwide.[5] And penicillin resistance was only
the kick-off to an arms race that continues to this day, and one
that we are not winning. Soon after researchers discover a new
antibiotic and doctors begin to prescribe it, bacteria resistant to
the antibiotic emerge. And they trigger a wave of resistance that
rapidly spreads through the world.

Antibiotic-resistant bugs already kill more than a million
people every year.[6] And because so many lives are at stake, we
urgently need to solve this problem. The path to a solution requires understanding. We need to understand how antibiotic
resistance originates, and how it spreads so rapidly. That's what
researchers like me study. And we already know a few things.
The spreading of antibiotic resistance is powered by the inces-
sant swapping and shuffling of genes – horizontal gene transfer
– that bacteria use to improve their lot in life. And it is acceler-
ated by modern travel, which allows people and their antibiotic-resistant bacteria to reach even the remotest places on earth
within mere days.

It follows that complete isolation from Western civilisation
– the isolation experienced by uncontacted Yanomami – should
protect against antibiotic resistance. Because uncontacted

Yanomami have never taken antibiotics nor been in contact with anybody who has, their microbiome should not be resistant to any modern antibiotic.

One would think. Because unfortunately, nothing could be further from the truth.

The stool and the skin of the Yanomami contacted in 2009 harboured bacteria whose genes conferred resistance against not just one, but eight different antibiotics. These included first-generation antibiotics like penicillin, but also more recently developed antibiotics that serve as drugs of last resort against drug-resistant bacteria.[7] The villagers may not have been contacted by other humans, but they surely had been contacted by antibiotic-resistant bacteria.

Perhaps these bacteria entered the village through indirect contact with the outside world? After all, the villagers did trade goods like arrows, machetes and clothing with other Yanomami closer to civilisation.[8] Or perhaps resistant bacteria got there by rain and wind, which can airlift bacteria across continents?[9] Could this account for their antibiotic resistance?

Perhaps, but the mystery does not end there. It deepens with other discoveries of antibiotic resistance, in places that are even further removed from our civilisation. Among them are sediments found deep underground, unearthed in 2009 from 170-metre-deep boreholes drilled in South Carolina and Washington state. These sediments are inhabited by bacteria, and when scientists tried to kill these bacteria with thirteen different kinds of antibiotics, they found that eighty-six percent of the bacteria were resistant to at least one antibiotic. More than sixty percent were resistant against more than one, and some were resistant against up to ten antibiotics.[10]

Even further removed from the planetary surface is a species of bacterium that lives on the bottom of the Pacific Ocean, more than one kilometre below sea level. It harbours a gene conferring resistance to four different antibiotics, including penicillin.[11]

Some bacteria are removed from humanity not just in space but in time, so far removed that it is easiest to detect them through the DNA they left behind. Among them are the deceased inhabitants of 30,000-year-old permafrost from northern Canada's Yukon territory. In these permanently frozen sediments, ancient bacterial DNA coexists several frigid metres below ground with DNA from extinct mammals like the mammoth. When scientists studied this ancient DNA, they found that it helps bacteria resist not just one but multiple antibiotics.[12]

Even longer lasting was the isolation of the Lechuguilla Cave in Carlsbad Caverns National Park of southern New Mexico. The cave has been leached out of the rock not by rain and rivers flowing on the surface, but by sulphuric acid that rose from deep in the earth's crust. Because of the cave's peculiar geology, it has been isolated from the surface for some four million years. Nonetheless, its microbial residents are well defended against our antimicrobial weapons. Seventy percent of them are resistant against three or more antibiotics, and one bacterium resists a whopping fourteen different antibiotics.[13]

When microbes live isolated from the planetary surface for a million years, or when they are buried deep underground, we can exclude even the most indirect contact with civilisation that the Yanomami have experienced. Such microbes cannot possibly have been transported from hospitals or patients by wind or water. They allow only one conclusion: their antibiotic resistance must be ancient, certainly older than the antibiotic era, and perhaps even older than humanity itself.

That may seem completely inexplicable, unless you know that many antibiotics are natural molecules. Among them is the archetypal penicillin. Alexander Fleming famously discovered it when a penicillin-producing fungus overgrew Fleming's bacterial cultures, using this chemical weapon to kill the bacteria. Fungi produce some natural antibiotics. Bacteria produce many others.[14] Both use these antibiotics in chemical warfare against their natural enemies.[15] Warfare is as widespread in microbes as in other organisms, such as the latex-producing plants we encountered in chapter 1. And such warfare triggers arms races, in which a defeated attacker develops new offensive tricks, such as the trick of plant-eating insects to cut a plant's latex canals. Likewise, the evolution of more lethal antibiotics triggers the evolution of better resistance. This evolutionary arms race between antibiotic-producers and resistance against them has been going on for countless million years. We humans are just the latest species to rush a crowded battlefield with the antibiotics we discovered.

But the natural origin of some antibiotics cannot explain *all* ancient antibiotic resistance. The reason is that some antibiotics are not natural. They are human-made. They include semi-synthetic antibiotics, natural molecules chemically altered by scientists to fool the defences of bacteria. They also include fully synthetic antibiotics, invented and manufactured in the laboratory. Remarkably, some bacteria are resistant against both kinds, including bacteria from uncontacted Yanomami and buried sediments. And these synthetic antibiotics lead to an even more surprising conclusion: bacteria must possess a dormant potential to combat molecules that they might not encounter for thousands of years, millions of years, or ever.

That is beyond inexplicable. It seems absurd. Is evolution clairvoyant, foreseeing that humans would eventually discover, synthesise and use non-natural antibiotics? Does evolution prepare bacteria even against antibiotics that may never be discovered? This possibility brings to mind the White Knight, a character in Lewis Carroll's novel *Through the Looking-Glass*. When the White Knight rescues Alice, the story's heroine, from the Red Knight, he tells her of his many inventions. They include a box whose lid is on the bottom to keep the rain out and a device for trapping mice should they appear on his horse's back.

'You see,' he went on after a pause, 'it's as well to be provided for *everything*. That's the reason the horse has all those anklets round his feet.'

'But what are they for?' Alice asked in a tone of great curiosity.

'To guard against the bites of sharks,' the Knight replied. 'It's an invention of my own.'[16]

Just like most horses do not encounter sharks, most bacteria will never encounter the synthetic drugs they are armoured against. Their defensive potential lies dormant. It is awakened only if we discover these drugs and go on the attack with them. It is hard to imagine that evolution has the kind of foresight needed to prepare bacteria against such eventualities.

The mystery, however, does have a solution. And the solution is quite mundane. It certainly does not require clairvoyance. But to understand it, we need to return to life's workhorse molecules, the long chains of amino acid building blocks called proteins. That's because proteins perform not just most jobs needed to keep life living. They are also responsible for antibiotic resistance.

I already mentioned that many proteins self-assemble – they *fold* – into convoluted three-dimensional shapes. Their amino acid strings are constantly jostled by countless other molecules that surround them and bounce into them. This jostling is caused by the molecular vibrations we call heat, which is also the engine driving protein-folding. In a protein's three-dimensional shape – also called a *fold* – some parts of the amino acid string that are far away in the string come to be near each other, just as in a skein of wool, where different parts of the same string lie next to each other. What keeps them together are chemical forces that attract some nearby amino acids to each other. Proteins with different amino acid strings fold into different shapes, and these shapes are responsible for the different and unique skills of many different proteins.

What I did not mention earlier is that protein strings never stop moving, even once they are folded. Although their shape is being held together by chemical attraction between nearby amino acids, they continue to be jostled by thermal vibrations, and so they keep wiggling and jiggling and vibrating. These motions are essential for what most proteins do, including the thousands of different enzymes that catalyse – accelerate – the chemical reactions powering all life on our planet. And some of these enzymes defend organisms by cleaving and destroying antibiotics.

To understand how molecular motions enable the accelerant magic of enzymes, consider what happens when an enzyme cleaves an antibiotic (or any other molecule). The antibiotic first needs to attach to the enzyme, and it does so in a location called a binding pocket. The amino acids near this pocket form a shape that is like a negative or a cast of the antibiotic. The antibiotic fits into this pocket a bit like a hand fits a glove, but unlike a hand whose muscles can help it wiggle into the glove, most molecules do not move on their own. Instead, they move only because other molecules bounce into them – heat. As an antibiotic molecule gets bounced around inside a cell, it may bump into the enzyme in various places before it hits the binding pocket in the right orientation. When that happens, the antibiotic stays there for a few moments, until it eventually gets dislodged again by further molecular bounces. Various forces can prolong this binding, for example the electrical attraction between a negatively charged antibiotic and a positively charged binding pocket.

During the brief time that the antibiotic remains attached to the enzyme, the enzyme continues to vibrate, and these vibrations are critical for what happens next. Because some amino acids in the enzyme's three-dimensional shape stick to one another, they cannot move freely, whereas others can move only in some directions, a bit like the blades of a scissor, which can move only in one plane, because they are linked by a pivot. Thus, even though heat vibrations buffet the enzyme from all sides, its amino acids' attractions and repulsions channel these vibrations in specific directions.

In an enzyme that cleaves an antibiotic, the enzyme's fold guides these motions to smash specific amino acids into the location where the molecule needs to be cleaved. The resulting collision splits the antibiotic, a bit like scissors cut through paper, except that the forces involved are chemical rather than mechanical, involving attractions and repulsions between atoms. The fragments eventually leave the catalytic site, dislodged by thermal vibrations. The enzyme continues to vibrate, ready to bind, cleave and destroy another antibiotic molecules.

Not all enzymes cleave molecules. Some join molecules or rearrange their atoms, but they too work by the same principle: a three-dimensional arrangement of amino acids channels thermal vibrations into guided motions that join molecules or rearrange their atoms. It's a simple principle, discovered time and again by evolution, and embodied in thousands of enzymes that catalyse countless chemical reactions. Each enzyme is a self-assembling nanomachine, an invisible marvel of the natural world.

Some of these machines are fast beyond imagination. One of them exists in the cells of our nervous system. This enzyme helps cells communicate. Its job is to help recycle neurotransmitters after they have ferried information between cells. A single molecule of this enzyme binds neurotransmitters, splits them, and ejects the debris at an astonishing speed of more than ten thousand molecules per second, accelerating one chemical reaction by a hundred-trillion-fold.[17]

Even faster is an enzyme that is responsible for the fizz of a soft drink. To create that fizz, the drink contains carbon dioxide, which forms carbonic acid as it comes in contact with water. When you open a bottle of soda, some of the carbonic acid molecules spontaneously fall apart. They release carbon dioxide, which forms small gas bubbles on the bottle's wall. This chemical reaction occurs without any accelerant, but it is slow. And that's a good thing. If the reaction was fast, your soda would explosively release all its carbon dioxide in a geyser of foam. Your saliva, however, contains an enzyme that accelerates this reaction. It is responsible for the fizzy feeling after you take a swig of a soft drink. The enzyme can cleave more than three hundred thousand molecules of carbonic acid per second and release their carbon dioxide all at once. (The real job of this enzyme is not to enhance your drinking pleasure, but to regulate the acidity of your stomach fluids and blood.)[18]

Enzymes may be fast but they are not perfect, and this is where I come back to ancient antibiotic resistance. Like scissors whose pivot is a bit loose, the guided motions of an enzyme can be sloppy. They may connect the wrong amino acids with a molecule bound to it. Or they may connect the right amino acid, but in the wrong place. Moreover, enzymes are also imperfect when binding molecules. Just like a small hand can slip into a large glove, so can some small molecules attach to a larger binding pocket. And just like a glove can accommodate a three-fingered hand or one with crooked fingers, some molecules do not have quite the right shape or electric charge, but they can still bind an enzyme.

For all these reasons, many enzymes that accelerate one kind of chemical reaction can also accelerate others.[19] Biochemists also call them *promiscuous* enzymes, because they catalyse reactions with multiple molecular partners.

A promiscuous enzyme usually has one favoured reaction, the one evolution has shaped it for. It accelerates this reaction massively. In addition, it also has multiple less-favoured reactions. It accelerates these reactions to a lesser extent, because its shape or motions are not quite right for them.

Many antibiotic resistance proteins are promiscuous, and this promiscuity helps solve the mystery of ancient antibiotic resistance.[20] It helps us understand why resistance even against human-made, synthetic antibiotics can be more ancient than humanity, and why such resistance can remain dormant for millennia. Just consider a family of promiscuous resistance proteins known as the beta-lactamases. Their name comes from *beta-lactam* – a ring of four atoms – and the suffix *-ase*, which designates enzymes that cleave molecules. In other words, beta-lactamases are enzymes that cleave and destroy a beta-lactam ring. This ring occurs in the famous – and naturally occurring – penicillin, but also in many other antibiotics, some natural, others synthetic. Different beta-lactamases work on different antibiotics with a beta-lactam ring, but – this is the important part – most destroy not just one antibiotic, but two, three, or many. Honed by natural selection, they fit one or few antibiotics with near perfection and destroy them rapidly. Other antibiotics they destroy more slowly.[21] Among them may be synthetic antibiotics that an enzyme would never encounter in the natural world.

Destroying antibiotics is not a bacterium's only means of surviving them. Another one is to pump toxic molecules out of a cell as fast as they get in. To this end, bacteria use specialised proteins called *efflux pumps* that get rid of a molecule before it harms them. Efflux pumps are complex protein nanomachines that connect a cell's interior to its exterior. They harbour cavities through which they move the molecules they transport. When such a molecule attaches to an efflux pump, the pump changes its shape to push the molecule through these cavities to the outside world, a bit like your gut contracts to push food through your body.[22]

Efflux pumps work on many kinds of molecules, not just antibiotics. For example, bacteria in our intestines use them to export bile acids, molecules produced by our liver to help us digest fats. Bile acids are highly toxic, so bacteria need to get rid of them. And the same pumps that help them get rid of bile acids can export multiple other kinds of molecules, including antibiotics. In other words, efflux pumps are highly promiscuous, exporting many molecules with the right size and shape. And this promiscuity also helps prepare the pumps for toxic molecules they may never have encountered.

Just like other promiscuous proteins, these pumps export ill-fitting molecules more slowly. If one such molecule is an antibiotic, enough of it may get left behind to kill a cell. But even in this kind of emergency, not all is lost. Bacteria still have a trick up their sleeve that illustrates how ingeniously evolution manoeuvres in the antibiotic arms race: bacteria can turbocharge their pumps to survive. Here is how.

Cells produce any protein by reading the DNA information of a gene – they *express* the gene encoding this protein. But they usually do not just express this information once, producing only one pump protein. They express it multiple times, producing multiple pumps.[23] Each pump finds its way to the membrane that separates the cell from the outside world and installs itself there – another marvellous example of molecular self-assembly. When a cell is flooded with antibiotics, these pumps become overloaded. But even when that happens, a cell can still help itself: it can crank up pump production. Ten times as many pumps will pump ten times as much antibiotic, which may be enough to survive.

This trick does much more than just help cells survive. It also helps researchers study what proteins do – any proteins, not just efflux pumps – and how widespread their dormant talents are. To do so, they engineer cells that *overexpress* a gene, and thus produce more of a protein than they normally would. To see how this could be useful, consider a promiscuous enzyme and one of the multiple reactions that it accelerates. Imagine that this reaction proceeds very slowly, and that the enzyme barely accelerates it. As a result, it will create only a tiny amount of the molecule that results from the reaction. This amount may be so minuscule that even our best scientific instruments cannot detect it. But when we overexpress this enzyme, making more of it, this amount may become detectable, and if so, a previously undetectable dormant talent of the enzyme has revealed itself. What is more, researchers can do this for more than one protein, such that different cells overproduce different proteins, and ask whether any one of these proteins can enhance survival.

A research team from New Zealand performed an experiment just like this, at an impressive scale: it created more than four thousand different kinds of engineered *E. coli* cells, such that each kind of cell overexpressed one of four thousand different genes. What the team discovered was just as impressive. Overproducing these proteins renders bacteria antibiotic-resistant, and not to just one, but to forty-one different antibiotics, some of them natural, others man-made. What's more, the bacteria also became resistant to forty-five additional toxins that can kill bacteria. In other words, the latent potential to resist toxins, caused by the imperfections of molecular machines, is enormous. The responsible proteins included familiar ones, like efflux pumps and toxin-destroying enzymes, but also proteins whose connection to resistance is not obvious. Among them was the very same protein that puts the fizz in your soda pop.[24]

This astounding resistance potential not only helps explain why bacteria in indigenous people and sealed million-year-old caves can be resistant to multiple antibiotics – they harbour enzymes and pumps whose promiscuity helps them disarm chemical weapons they have never seen – but also helps explain how antibiotic resistance that is strong – the kind that kills vulnerable patients – can arise rapidly from resistance that is undetectably weak. Bacteria can evolve such antibiotic resistance almost instantly, much faster than the few years it may take for new resistant bugs to spread through clinics worldwide. In fact, antibiotic resistance can evolve so fast that we can even observe its evolution in real time, in the laboratory. That's what researchers like those in my Zürich lab learn when we challenge bacteria to survive ever-increasing amounts of antibiotics.

One of these researchers is Shraddha Karve, a former Indian postdoctoral researcher in my lab. I hired Shraddha because she was deeply interested in how bacteria survive challenging environments that contain various poisons, such as antibiotics and heavy metals, which inactivate essential enzymes. She had already performed evolution experiments as a graduate student in India, and from this earlier work, I knew that she had a knack for designing simple yet informative experiments. An experiment is a way of asking nature a question, but often the answer that nature gives us is vague, unclear or ambiguous. The best experimenters know how to pose the question in a way that forces nature to give a clear and concrete answer, like a skilled trial lawyer who will extract exactly the information they need when examining a witness. This is a highly prized skill – many aspiring scientists never acquire it – and Shraddha had it in abundance. She put this skill to use in an experiment that taught us how fast evolution can evolve new antibiotic resistance, but not just that. It also taught us that evolution can quickly create organisms with multiple new and dormant talents.

Shraddha began by exposing cells of the bacterium *E. coli* to a small amount of an antibiotic called trimethoprim, an amount that was not sufficient to kill the bacteria. Over the next twenty-four days, she steadily increased the amount. Twenty-four days is little time for evolution, but even during this brief period, the cells evolved to thrive on a thousand-fold greater amount of the antibiotic. This part of Shraddha's experiment illustrates how fast – in less than four weeks – evolution can massively increase antibiotic resistance. But it was only the first part. The second part was more important. Shraddha placed her evolved bacteria into ninety-five different hostile environments in which their ancestors – the cells that she had started with – had not been able to survive. Some of these environments contained antibiotics, but others contained different poisons, such as heavy metals, detergents, oxidants and toxic salts. Then she asked whether the cells could survive in any of these environments. That may seem like an odd question. The bacteria had never lived in these environments, so why should evolution have helped the bacteria survive? But that's exactly what Shraddha found. The evolved bacteria had become able to survive in sixteen new environments they had not lived in, and that would have killed their ancestors.

The outcome of this experiment could have been a fluke, a one-off freak observation that was unusual, and that we would never see again. So we did what scientists always do to avoid that possibility. We repeated the experiment. In fact, we repeated the experiment in a slightly different way to obtain additional information. Specifically, Shraddha repeated the experiment four times, each time selecting for resistance to a different antibiotic. And each time the outcome was the same. First, within a few weeks, the bacteria evolved massive resistance to the antibiotic. And second, they also became able to survive in multiple other toxic environments that they had not previously encountered.

Shraddha could have stopped there, but if she had we might never know why that was possible. We might have been left with the possibility that evolution is clairvoyant. So we decided that Shraddha should sequence the entire genome – some 4.5 million DNA letters – of her evolved bacteria. By doing so, we hoped to identify DNA mutations that could explain not just their increased antibiotic resistance, but also their survival in other new environments they had not encountered. And our hopes were fulfilled. Each genome had acquired DNA mutations, and some of these mutations altered proteins known to be promiscuous and involved in resistance to toxins. For example, some mutations cranked up the production of notoriously promiscuous efflux pumps. Clairvoyance was not required. Promiscuity was enough.[25]

All this is bad news for medicine. Not only does antibiotic resistance evolve almost instantly, the DNA mutations that increase resistance also enable bacteria to survive toxins – including antibiotics – they have never before encountered. In other words, we are part of an arms race that we cannot win decisively. Bacteria may already harbour dormant resistance against any new antibiotic we deploy. If they do not, they can rapidly evolve this resistance, and in doing so, create the ability to survive even more antibiotics, ones that we have not even discovered. New antibiotics not only awaken slumbering microbial talents, they help create more such talents. Nature's potential to innovate in this arms race seems bottomless.

If this sounds gloomy, it is worth keeping in mind that we need not be losing the race either, at least not permanently. Newly developed antibiotics do work for a while, until resistance against them has spread. Their useful life can be prolonged if doctors prescribe them only when needed. That can buy us time to develop further antibiotics or altogether new strategies, such as to kill bacteria with viruses. If we use our ingenuity and resources wisely, we need not concede the arms race to microbes. But in the long run, we will have to learn to coexist with them.

* * *

For an organism to thrive, an environment must not contain lethal poisons. But that's not enough. The environment must also supply nutrients that provide the organism with energy and essential building materials. These nutrients are digested by metabolic enzymes that obey the same chemical laws as antibiotic resistance proteins. In particular, they are also subject to the causes of promiscuity – sloppy molecular motions. That's why promiscuity and its consequences extend far beyond drug resistance. They affect all of metabolism, the complex road network of chemical reactions that keeps life living. And they are the reason why metabolism also harbours a vast dormant potential, the ability to use foods that an organism has never encountered.

I come back here to *E. coli*, because it's been studied for more than a century and by thousands of researchers.[26] What they discovered about the promiscuity of its enzymes is remarkable: more than three hundred of the thousand plus enzymes that keep the bacterium *E. coli* alive are promiscuous.[27] And this estimate may be too low, because our knowledge about *E. coli* is still incomplete – even this tiny living bag of molecules is complex beyond imagination. Many more enzymes than we know may be promiscuous, and the promiscuous ones may catalyse many still unknown reactions. But even what we know today tells us that *E. coli*'s metabolic promiscuity has remarkable consequences. For example, it allows *E. coli* to digest nineteen different 'foods' on which it would otherwise starve to death, because some of its promiscuous enzymes can help create useful nutrients from useless molecules.[28]

The power of that metabolic promiscuity is highlighted by another remarkable experiment, this one performed at Oxford University by Macarena Toll-Riera, also a former postdoctoral researcher in my laboratory.

Macarena worked with ninety-five populations of bacteria, allowing each population to evolve in a different environment. Any one environment harboured only one kind of nutrient, and different environments harboured a different nutrient.[29] At the beginning of her experiments, her bacteria were able to survive in some environments, but not in others. By the end, the bacteria had experienced DNA changes in their genome that allowed them to adapt to all but four of these environments. They grew and divided faster in those environments where they had already been able to survive, and they survived in environments where they would previously have starved to death. All that evolution took no more than thirty days.

That's already a remarkable discovery, but what came next was even more remarkable. Macarena asked whether each of her ninety-five evolved bacteria also survived better in the ninety-four other environments that they had not experienced during evolution. Once again, this is an odd question – none of these bacteria had ever lived in these environments, so how could they possibly thrive there? But it's exactly what might happen when enzymes possess slumbering talents. Indeed, the evolved bacteria grew better in new environments, and not in just one, but on average in sixteen environments they had not lived in.[30]

Experiments like this show how fast evolution can improve an organism's lot and rouse its dormant potential, but they do not plumb the true depth of this potential. They do not ask how many different molecular partners a promiscuous enzyme can really have. To find out, a different kind of experiment is needed. In this experiment, one exposes an enzyme to many different molecules and asks whether the enzyme can help few, all, or most of these molecules react.

One such experiment focused on especially important and widespread enzymes that transfer phosphates between different molecules. Phosphate – a phosphorus atom bound to three oxygen atoms – is everywhere in living cells. Every single one of our DNA building blocks contains phosphate. Life's universal energy carrier ATP has three phosphates. Many proteins can be switched on or off by attaching phosphate to them and so on. That's why reactions involving phosphate are important, and why organisms produce not just one but many enzymes that transfer phosphates between molecules.

In a 2015 study researchers examined two hundred such enzymes, exposing each one of them to 167 different phosphate-containing molecules. For every enzyme-molecule pair – that's more than sixteen thousand pairs – they asked whether the enzyme could catalyse a reaction with the molecule. The answer was an emphatic yes. More than half of the enzymes catalysed reactions with 6 to 40 molecules, and 50 enzymes catalysed reactions with up to 143 molecules.[31]

If each of a bacterium's thousand or so enzymes can help catalyse some five reactions – a modest estimate given experiments like this – then its metabolism can accelerate five thousand reactions, five times more than without promiscuity, and many more than it would need to survive in any one place.

This metabolic potential can even extend to otherwise poisonous molecules. I have already mentioned the ability of bacteria to survive antibiotics; some bacteria can do much more than that – they can use antibiotics as nutrients. The atomic bonds of antibiotics contain a lot of energy and carbon to build new cells, if only cells could harvest these resources instead of getting killed. And indeed, some cells can. In a 2008 study, researchers force-fed bacteria isolated from soils around the world the very same antibiotics doctors use to kill them. And astonishingly, each soil harboured bacteria that could not only survive antibiotics, but use them as their only source of energy and carbon. These bacteria came from pristine soils with minimal human contact, and some of the antibiotics do not even occur in nature. Bacteria's latent metabolic potential can turn even poisons into foods.[32]

Gene transfer between bacteria further enhances this metabolic potential. I mentioned in the previous chapter that transferred genes – and the enzymes they encode – create not only new roads in a metabolic highway network, but also new links between existing roads.[33] The new metabolic traffic created by any one such enzyme can help a cell survive in multiple new environments, even if that enzyme is not promiscuous. A promiscuous enzyme enhances this potential even further, because it can connect multiple metabolic roads.

Cells are microscopic bundles of innovation potential so great that we are only beginning to understand this potential. It does not even stop with metabolism, because accelerating chemical reactions is only one among many tasks that are essential to keep a cell alive. Another one is to turn genes on or off. Enzymes, it turns out, can do that too. They are much more than just flexible catalysts.

When a cell expresses the information in a gene to make a protein, it first transcribes an RNA copy of the gene. Specialised proteins called *transcriptional regulators* govern this process – they *regulate* transcription. These regulators can turn a gene's transcription on or off. Not only that, they also regulate how many RNA copies are made – the more RNA copies, the more proteins a cell can translate.

A regulator works by attaching to a short stretch of DNA near the beginning of the gene. This attachment signals to a cell that it can start transcribing.[34] Cells harbour dozens to hundreds of these regulators. Each regulates a different set of genes. Some regulate genes responsible for digesting nutritious sugars, others regulate genes that protect cells from excessive heat, and still others regulate genes that repair damaged DNA.

What it takes to be a regulator – bind DNA, control transcription – is different from what it takes to be an enzyme. Surprisingly though, some enzymes double as regulators. One of them is an enzyme that helps metabolism build the amino acid arginine. The same enzyme is also capable of regulating genes that have nothing to do with building arginine – they are needed when cells use oxygen to produce energy.[35]

Let me pause to underline how remarkable this is. If a promiscuous enzyme is like a virtuoso pianist who can also play the guitar, then this kind of protein is like a musician who is also an expert aircraft engineer or a doctor. Its range of skills extend far beyond what evolution has trained it for.

Biochemists call such proteins moonlighting proteins, because they have a day job and one or more side jobs that require more diverse skills than those of enzymes. But don't think of these side jobs as mere hobbies. They can be essential for survival.

We know of hundreds of such multitalented proteins, but the most astounding ones occur in the eyes of humans and other animals, because they help us see.[36]

The lenses of our eyes focus light by changing their shape depending on whether we focus on a nearby or distant object. The lens material changes the direction of light rays, like water refracts light that enters it from the air. To refract light, a lens needs to be denser than the medium surrounding it, which is accomplished by proteins – crystallins – that the lens harbours in very high concentrations. In contrast to many other proteins, which form opaque clumps when highly concentrated, crystallins remain transparent in the lens. (They clump only when damaged, for example by UV light. And when they do, our lenses form cataracts – they become cloudy and we go blind.)

Even more remarkable than the transparency of crystallins is where they come from. They do not occur in the eye alone, but in many other organs in our body, where they have a different job: they catalyse chemical reactions. In other words, crystallins are enzymes in our metabolism. Some of these catalysts are so ancient that they predate the evolution of eyes. But when eyes evolved, evolution changed the regulation of these enzymes so that the eye would produce lots of them. The job of a crystallin is a moonlighting job. Its evolution required proteins whose slumbering talent is especially simple: they remain transparent when overproduced.[37]

Animals and bacteria are not alone in harbouring molecules with untold potential. Plants do too. Because plants cannot run away from the many enemies that feed on their roots, leaves or other body parts, they had to evolve a special superpower – chemical defence. Think of it as a sophisticated form of passive aggression. And multitalented molecules, it turns out, play a key role in it.

One example of this superpower is embodied in the weaponised sticky plant secretion we heard about earlier – latex and resins. But they are only two armaments in plants' vast chemical war-chest. Others include *cyanogenic* molecules. The name says it all. When an animal takes a bite out of a plant containing such a molecule, the molecule releases cyanide. That's the same lethal poison the Nazis used to murder prisoners in the gas chambers of Auschwitz. Cassava or manioc tubers – staple foods in Africa and South America – contain such molecules, and that's why eating them will get you cyanide poisoning unless the tubers are cooked or soaked. Plants are prolific inventors when it comes to these cyanide-releasing poisons, having evolved sixty different kinds of them.[38]

Even more diverse – and no less devious – are molecules that resemble animal hormones. The best-known of them is ecdysone, a hormone that helps insects shed and replace their hard chitin armour as they grow. It also helps transform caterpillars into butterflies. Plants produce their own variants of these molecules – more than two hundred we know of and counting – to mess with the growth and development of insects that feed on them.[39]

Tannins make your mouth shrivel when you bite into unripe fruit. They are chemical weapons aimed at your digestive tract, where they bind your digestive enzymes and plant proteins to make them harder to digest. They belong to a gigantic family of defence chemicals with more than nine thousand members. But even their number is dwarfed by the twelve thousand alkaloids. We know some of them as recreational drugs like nicotine and caffeine, but really they are chemical defences against herbivores. Nicotine, for example, is highly toxic to herbivores, because it disrupts their neural communication.[40]

All in all, plants produce two hundred thousand different kinds of molecules that we know, and probably many more that we don't. Most of these molecules are not essential to plant life, but they serve specialised roles, such as in chemical warfare.[41] I mention this chemical diversity, because it would not exist without promiscuous enzymes. But the enzymes behind it are not the kinds of metabolic enzymes that extract energy from nutrients. They are *biosynthetic* enzymes that build more complex molecules from simpler ones. And a promiscuous biosynthetic enzyme catalyses multiple reactions that produce multiple complex molecules. Such enzymes help a plant produce not just one defence chemical but complex cocktails of them. For example, an enzyme in tobacco plants helps build a molecule that wards off fungal infections, and this enzyme produces no fewer than twenty-four other molecules as by-products.[42]

Especially potent in this regard are enzymes called cyto-chromes P450, because some of them catalyse hundreds of different reactions.[43] What is more, the biosphere contains more than fifty thousand different variants of these enzymes, which differ in the spectrum of reactions they catalyse. Many species produce more than one cytochrome, and plant species often produce many. Rice alone, for example, has more than four hundred different variants of these promiscuous enzymes.[44] Promiscuity like this goes a long way towards explaining the diversity of chemicals in plants.

Only some of these chemicals are useful. The rest may be as useless as the inventions of the White Knight, at least until a plant encounters the right environment. That's when they can become potent weapons. In fact, a new environment is all that's needed to transform some plants into destructive raiders that wipe out everything in their path. They are also known as invasive plants, and they can wreak havoc on the environments they invade.[45]

Among these species is one that has plagued the United States for almost a century. Its name is kudzu, a vine that climbs and spreads aggressively. It came to the US from Asia in the late 1800s, and was used to combat soil erosion in the dust bowl era of the 1930s.[46] Since then, however, it has become a noxious weed. Widespread in the southern US, it quickly over-grows everything in its path – not just other vegetation, but also abandoned cars and dilapidated buildings, covering them with a suffocating mass of green foliage. Kudzu grows so fast that US troops used its dense blanket of leaves to camouflage their equipment on Pacific islands in World War II.

As in land, so on water. The water hyacinth was first intro-duced in the 1800s from South America as an ornamental pond plant, not the least because of its showy pale purple flowers. More striking though is how fast it reproduces and spreads, ei-ther through the three thousand seeds a plant releases annually, or through its network of runners that continually sprout new plants. As a result, water hyacinths grow amazingly fast – twelve acres a day have been reported from African Lake Victoria. They can quickly cover entire waterways with a mat of tissue dense enough to walk on. This mat not only starves other plants of light, it also kills fish by preventing air from mixing with water. It clogs irrigation channels, blocks shipping routes, entangles the propellers of fishing boats and obstructs the water intake of hydroelectric power plants. In the US alone, it causes some 120 billion dollars worth of damage per year.

Remarkably, invasive plants like these are usually much less destructive in their native habitat. They spiral out of control only as they enter new territory. One reason is a lack of natural enemies. For example, the weevils and moths that feed on water hyacinths, keeping them in check in their native South America, do not occur elsewhere. That's why releasing these insects in a new territory can help slow down an invasion.[47]

But liberation from natural enemies is not the only reason why invaders steamroll the native flora. Many invading plants also produce chemicals that blur the line between defence and offence. Some of them stop the seeds of other species from germinating. Others shrivel their competitors' roots. Still others are even more devious. These chemicals mask the attractive scent of flowers from competing plant species. In doing so, they lead the insects that pollinate the flowers of their competitors astray, and thus prevent the competitors' reproduction.[48]

Some of these chemicals are especially devastating for plants in the new territory. Whereas plants in the old territory had time to adapt to them and can thus coexist with the aggressor, plants in the new territory may never have experienced these chemicals. As a result, they are defenceless against them.

Among the invasive species that use such chemical weapons are pine trees. Mostly native to the northern hemisphere, they have successfully invaded southern Africa, Australia and South America, where they are threatening native species.[49] The needles they shed release chemicals that prevent other plants from growing – that's why the ground beneath a stand of pine trees is often barren, even though it may be well-lit and moist.[50]

Another example is the spotted knapweed, an innocuous looking plant whose branched stem grows up to a metre in height and terminates in feathery purple flowers. Since it invaded the United States from Europe, it has spread over more than 28,000 square kilometres. The US department of agriculture calls it a noxious weed, because it suppresses the growth of plants around it, including those preferred by grazing cattle. It produces its own herbicide, a molecule called catechin that prevents nearby seeds from germinating.[51]

Yet another invasive species has the dubious distinction of being one of the ten worst weeds in the world. It is the bitter vine (*Mikania micrantha*), a native of South and North America that has invaded Asia, where it is spreading rapidly, partly thanks to the cocktail of seventy-nine different chemicals it produces. When extracted from its roots and leaves, these chemicals can inhibit the growth of twenty different tree species.[52]

Some members of such a chemical cocktail may have been useful already in a plant's native territory. Others may be useless by-products of promiscuous enzymes. Still others may be useless at home, but become potent weapons abroad. These are the molecules whose potential is dormant. In the right environment, this potential can erupt as a devastating species invasion.

* * *

The idea that evolution endows the old with new uses has deep roots. It goes back at least to Darwin's *Origin of Species*, where he mentioned the swim-bladders of lung fish as examples.[53] These gas-filled sacs help fish to control buoyancy and stay at their current water depth without expending energy, a bit like the ballast tanks of a submarine.[54] They originated as gut appendages, but much later, when fish began to conquer land, they would prove useful for a reason unrelated to their buoyancy control: they help extract oxygen from the air. We owe our air-breathing existence to the latent potential of a swim bladder to become a lung.

Legions of similar examples are the stuff of secondary school biology: the jaw bones of reptiles turned into three tiny bones that transmit sound from our eardrum to our inner ear.[55] The feathers that keep birds aloft originally served to keep them warm or dry. The sharp beak of the Kea, a parrot from New Zealand, helps it feed on seeds, but when sheep were introduced to New Zealand, it also allowed the Kea to slice through the sheep's skin and feed on their flesh.[56]

For a long time, traits like swim bladders were called preadaptations, because they are more than just adaptations. They were well suited for a job like breathing air even before the job existed. However, many biologists eye that word sceptically. It suggests – falsely – that evolution has foresight, as if something inside our fish ancestors knew that breathing air would one day become a hot new skill.

It was not until 1982 that biologists could start to avoid that word. That's when palaeontologists Stephen Jay Gould and Elisabeth Vrba invented a new and fancier one: exaptation. Their neologism stuck, perhaps *because* it was new and therefore less loaded. Since then we say that middle ear bones are exaptations for hearing – exapted from jaw bones – feathers are exaptations for flying and so on.[57]

When Gould and Vrba coined this word, biology had already entered the molecular era that began with the discovery of DNA's double helix in 1953. It soon became clear that the molecules of life were abundant sources of exaptations, perhaps even more so than large structures like feathers and bones. For example, to produce the milk sugar lactose, mammals like us need a protein whose origins have little to do with milk. It evolved from another protein that kills bacteria in our saliva. This protein – lysozyme – is ancient, older than mammals, and possibly older than animals, because some bacteria also produce a version of it. Some viruses even employ it to kill bacteria.[58]

Other examples include the moonlighting crystallins that were exapted from metabolic enzymes, as well as some proteins that prevent the blood of Antarctic fish from freezing, which are related to enzymes in the pancreas.[59] In all these exaptations, something new and useful originated from something old and useful.

Promiscuous proteins break this mould. On the body of an uncontacted Yanomami, a protein that destroys a twenty-first-century antibiotic is useless. So is a promiscuous enzyme's ability to produce a defence chemical without a matching enemy. They are solutions in search of a problem, by-products of whatever else these enzymes are doing. In this sense, they are like the inventions of the White Knight. When they become useful – with the right antibiotic or the right enemy – they become exaptations, but not the ones from biology textbooks. They do not merely transform useful things into other useful things. They transform the useless into the useful, and they do so without the rare and precious DNA mutations that improve proteins. All they need is a changing environment.[60]

Because so many – some argue all – enzymes are promiscuous, every single cell alive today harbours thousands of molecules with such untapped potential.[61] They are the sleeping beauties of the protein world that keeps life going. In this world, White Knight inventions may outnumber adaptations, just like bacterial enzymes may catalyse many times more reactions than they need to survive. Nature's potential to innovate may far exceed its need for innovations at any one time. Neither Darwin nor Gould had any idea about the vastness of this untapped potential. Nor did they know that underneath it lies another layer of biological innovation that is even more fertile.

5

The birth of genes

A junkyard contains all the bits and pieces of a Boeing 747, dismembered and in disarray. A whirlwind happens to blow through the yard. What is the chance that after its passage a fully assembled 747, ready to fly, will be found standing there? So small as to be negligible, even if a tornado were to blow through enough junkyards to fill the whole Universe.[1]

THESE LINES COME FROM TWENTIETH-CENTURY British astronomer Fred Hoyle, who used them to argue for 'panspermia', the idea that life did not originate on this planet, but was seeded from outer space. They are also known as Hoyle's fallacy. What's wrong with them is the idea that the Boeing 747 must be assembled in a single event. No respectable biologist today believes that evolution works this way. Evolution does not proceed in big leaps, but in smaller steps. Each step must be preserved – the key task of natural selection – before the next step can be taken.

Crackpot ideas aside, biologists have been debating since Darwin's time how small evolution's steps really are. Today the predominant view is that most steps are so tiny as to be effectively imperceptible. This view is called gradualism and Darwin himself shared it. Gradualism is supported by the slow change of many species in evolutionary time. It also fits with what we know about DNA mutations and how they modify genomes. Because many mutations alter only one of the million DNA letters in a genome, they can thus change only one among thousands of proteins at a time. Most such mutations will affect a whole organism very little.

Gradualist thinking also applies to how evolution creates new features of life. Take the origin of new proteins and new genes – stretches of DNA nucleotides that a cell transcribes into the carbon copy of RNA, which it then translates into the amino acid string of a protein. Here is what a leading expert has written about the origin of new proteins and genes:

'The probability that a functional protein would appear *de novo* by random association of amino acids is practically zero. In organisms...creation of entirely new nucleotide sequences could not be of any importance...'[2]

And this was not just any expert. It was French Nobel laureate François Jacob, a luminary of twentieth-century genetics who taught us how cells determine when a gene is transcribed and translated.

Jacob's lines are part of a celebrated 1977 essay on a broader gradualist argument that is widely accepted today: evolution is like a tinkerer with a huge workshop full of junk, devices in various states of assembly and repair, gizmos with half-forgotten uses, and countless tools just as likely to be working as to be broken. And like a tinkerer, evolution modifies, fiddles and plays with these parts, assembling them into ever-new contraptions, gadgets and molecular machines. In other words, evolution modifies life's existing parts to create new parts – wings from arms, lungs from swim bladders and so on – rather than creating radically new things – including new genes and proteins – in one giant leap.

Today, many years and thousands of sequenced genomes later, we know that Jacob's view of evolution as a tinkerer is largely correct. But Jacob was also dead wrong about something equally important. New genes need not originate from old ones. They can and frequently do appear from scratch, so frequently that biologists coined a special term for such genes: *de novo* genes.

Jacob did not know this. The discovery had to await the genomic era, which began only in the twilight years of Jacob's life. The existence of *de novo* genes proves that evolution *can* create complex features of a genome from scratch. And they teach us that we may not give evolution's creative powers enough credit. *De novo* genes are the subject of this chapter, and not just because they highlight this power. They illustrate once again that evolution often creates more than it needs. That's because most *de novo* genes are created dormant. They emerge long before they might help an organism survive.

Before I say more about *de novo* genes, it is useful to understand how evolution creates new genes the old-fashioned way, by tinkering with old genes. To get us there, I need to say a few more words about DNA mutations.

I already mentioned that most DNA mutations are caused by the protein repair crews that tirelessly maintain the DNA in our cells. They mend damage from high energy radiation, and from the toxic chemicals produced by our own metabolism. Some of them fix individual mutated DNA letters, while others fix entire text strings, patching them up with text from elsewhere in the genome.

These repair proteins are great molecular machines, but they are fallible. Sometimes their proofreading overlooks a single mutated DNA letter, a point mutation. Such a mutation can kill, for example when it destroys the information needed to make an essential protein, like a typo that garbles a word essential to understand a text. Some unrepaired mutations, however, do little harm, for example when they occur outside a gene. Harmless mutations can get passed on to the next generation, and become permanently preserved in an evolving genome. They are important for the molecular clocks I mentioned in chapter 2.

DNA repair crews also make another kind of mistake. While copying DNA from elsewhere in the genome to fix a text passage, they occasionally copy-paste text into the wrong place. The result is a DNA *duplication*, which can comprise a few, thousands or millions of DNA letters. When the duplicated DNA includes a gene, the gene itself has become duplicated.

An original gene and its duplicate are initially identical to one another. With the passing of time, however, both genes experience the same drizzle of point mutations that rains down on all of our genome. Many of these mutations are repaired but some fall through the cracks. Among them are not only harmful mutations that kill their bearer, but also harmless mutations that survive and are passed on from generation to generation. Little by little, these surviving mutations cause the two genes to *diverge* – their DNA texts gradually become less and less similar.

Duplicate genes often diverge especially rapidly, because most mutations that affect them fall into the harmless category.[3] It's not hard to understand why. Even the most damaging kind of mutation, one that inactivates one of the two genes and destroys its ability to make a useful protein, can be harmless, as long as the other, redundant copy is still around.[4] In fact, such destruction is the fate of most gene duplicates. It affects ninety percent of them in some species, and is caused by one of those mutations that garble the molecular meaning of a duplicate, which is then no longer transcribed or translated into a useful protein.[5] Such a mutation creates a *pseudogene*, and it leaves the other copy as alone as it was before the duplication in performing its tasks.

This process – duplication, mutation, divergence and eventual destruction – can repeat itself over and over, creating not just pairs but entire families of duplicates. Any one family can comprise hundreds of members, some of them active genes, many others pseudogenes. And like human families whose members resemble each other physically, these family members usually resemble each other in their DNA text. The most recently duplicated members are most similar. Older duplicates are less similar. The oldest duplicates may have mutated so often that they no longer resemble their kin.

Most DNA duplicates passively, through the errors of DNA repair enzymes, but some DNA has a special talent. It is mobile. It can copy-paste itself to new places in the genome, because it harbours genes that help it move. Although this ability does not shield mobile DNA against the ceaseless drizzle of point mutations, such mutations are even less harmful in mobile DNA than in other duplicated DNA. That's because the genes of mobile DNA are prime examples of what biologist Richard Dawkins calls selfish genes: they are genomic parasites whose highest purpose is to go forth and multiply within a genome.[6] They do not serve the greater good of keeping us alive. For this reason, mutations that cripple them go unpunished by natural selection. As a consequence, our genome also harbours previously mobile DNA that mutations have paralysed and rendered immobile, lots of it, thousands of crippled copies.

DNA duplications that arise passively through botched DNA repair or actively through mobile DNA are frequent. They can be more frequent than the better-known point mutations that change single DNA letters.[7] They are so frequent that more than half of the three billion letters in our genome are remnants of past DNA duplications, some very recent, others hundreds of millions of years old. Many of these remnants belong in gene families whose members still resemble each other. Many others – especially pseudogenes – have diverged beyond all recognition, through countless mutations that have randomised their

DNA sequence. They are like footprints that have been erased by a steady rain, just more slowly, over millions of years.

Fortunately, the story of duplicate genes is not just one of erosion and destruction, otherwise evolution might not have gotten very far. Just like a tinkerer who occasionally stumbles upon a useful gadget, evolution occasionally creates innovative genes through a combination of duplications and point mutations. Such mutations help create proteins that can ferry new kinds of nutrients into a cell, molecular motors that can move heavier loads, or enzymes that catalyse new chemical reactions. I already mentioned the cytochrome P450 enzymes. In our genomes, they are encoded by a family of more than fifty duplicated genes, which help us synthesise vitamins and hormones, but also destroy toxic molecules in our liver. But the size of this gene family in humans is dwarfed by an even bigger clan in plants like rice, where the family has expanded to hundreds of members. They help plants create a potent chemical defence cocktail. Each member is a gene duplicate that survived multiple mutations. Not only that, these very mutations have endowed it with new chemical skills.[8]

This is how Jacob imagined that all new genes evolve.[9] Mutations that duplicate and then modify DNA often create the useless junk of pseudogenes, but occasionally they hit the jackpot, a mutated DNA sequence that encodes a new kind of protein. These jackpots are rare, but over the eons, they add up: our genomes are filled with thousands of duplicate genes whose new skills emerged when nature modified the skills of their ancestors.

Jacob deserves a lot of credit for this vision, because he had no way of knowing whether it was true. In 1977, the technology to read our genome's DNA and identify its genes was in its infancy. Only in the 1990s – the beginnings of the genomic era – would the first sequenced genomes prove the truth of his vision. Or part of it, I should say. Because they also taught us that

Jacob missed a burbling source of new genes, random stretches of DNA that encode nothing.

One can hardly blame Jacob for arguing against *de novo* genes, because a gene really is a very special stretch of DNA. It has complex features that Jacob knew well – he had discovered some of them – and whose origin is hard to imagine. Some of these features allow a cell to transcribe the gene's DNA into an RNA copy. Transcription is the job of an enzyme called RNA polymerase, which needs help from a protein we already heard about – a transcriptional regulator. Such a regulator binds specific DNA words near the gene and directs the enzyme to start transcribing. The enzyme cannot transcribe unless such a word – between six and twenty letters long – happens to exist near the gene.

The transcript then must be translated into protein, which requires not just one protein like RNA polymerase, but a complicated biochemical machinery called the ribosome. This ribosome scans the transcript from one end until it finds a specific letter triplet that spells AUG. (The letter U in the transcript stands for uracil. It corresponds to the DNA letter T for thymine, which the polymerase transcribes into the RNA letter U.) This AUG triplet is a signal to the ribosome to start translating, and the ribosome translates it into a specific amino acid called methionine.

From there on, the ribosome moves along the transcript, scans every successive letter triplet and translates it into one of twenty amino acids, using a genetic translation code that is almost as old as life itself. As the ribosome moves along, it links the amino acids to form a continuous protein string, until it encounters one of the three triplets UGA, UAA or UAG. To the ribosome these triplets have a specific meaning. They embody the command 'stop translating'. Together, the start signals, the stop signals and the letter triplets in between form an *open reading frame* or ORF, a shorthand for a DNA region or its transcript that encodes a protein.[10]

To qualify as a gene, a stretch of DNA must not only harbour specific DNA words that guide transcription. It must also contain an open reading frame that is translated into protein. Given these requirements, it's hard to imagine indeed that a gene could emerge from scratch.[11]

Hard to imagine, but still true. The first clues arrived early in the genome era, after the first genomes of iconic species like humans, fruit flies and *E. coli* had been sequenced. At the time, it had already become clear that reading multiple genomes would be much more useful than reading just one. Take our own genome. Its DNA was first sequenced in the year 2001, which taught us a lot about human biology. It taught us how many genes we have, which proteins they encode, and how our metabolism works. This knowledge was illuminating, to be sure. But it was like a dim candle compared to the bright sunlight of insight we can get by comparing our genome to that of other organisms, such as that of the chimpanzee, which was sequenced in 2005. For such genome comparisons, one needs to first *align* two genomes – line them up side-by side, letter by letter – and then compare their features, most notably their genes.

The human and chimp lineage split from a common ancestor some five million years ago. Since then both lineages have evolved separately and experienced different mutations. Natural selection eliminated harmful mutations and allowed harmless mutations to survive. The most interesting surviving mutations are the innovative ones, those that make humans human. One can find them by identifying mutations that are unique to the human lineage. Among them are mutations that have helped humans produce the wide range of sounds needed for languages. Other mutations helped turn our hands into sophisticated tools, our brains into unrivalled data processors, and our bodies into the high endurance running machines that helped our ancestors survive in the African savannah.[12] More generally, comparative genomics – the science of comparing genomes – can teach us what makes species unique and what makes them similar.

Comparative genomics also reveals that natural selection has tolerated many mutations in some genes, but it has tolerated few or no mutations in others. For millions of years, these genes have been preserved unchanged – the technical term is *conserved* – because they play such an important role in sustaining life that any mutation in them is punished by death or extinction. This is a general principle of genome evolution: if a gene's letter sequence changes slowly in evolution, it means that the gene is important, because natural selection eliminates most DNA mutations in it. Comparative genomics is crucial to identify important genes.

Each newly sequenced primate genome – from chimps to gorillas, orangutans, macaques and others – taught us a bit more about the common biology of all primate species and the unique biology of each one of them. Unfortunately, sequencing the typical primate genome is a daunting challenge, because our genomes are so huge. Their almost three billion letters would amount to three million pages of single-spaced text on standard letter paper.[13] The genomes of other species are smaller and therefore easier to read. That's why in 2007, merely two years after sequencing the chimp genome, researchers had already read more than a dozen genomes from various *Drosophila* fruit fly species. These genomes comprise only two hundred million letters, ten times fewer than the human genome.[14] Sequencing the genomes of bacteria is easier still, because they comprise only a few million letters, about one thousandth of the human genome. That's why genomic databases are filled with thousands of genomes from bacteria like *E. coli*.

Comparing an ever-increasing number of genomes resulted in many discoveries, but none more mystifying than this one: every newly sequenced genome contained hundreds to thousands of genes whose DNA was unique, bearing no resemblance to DNA in any other organism. Such genes were called *orphans*.

Orphans are good candidates for *de novo* genes.[15] I say candidates, because in the early days of genomics, researchers could not exclude a more mundane explanation of their origins. The reason is that few genomes had been sequenced at the time, and most of these came from very distantly related organisms. Among them were two species of yeast, including the familiar brewer's yeast, with hundreds of orphans each. The most recent common ancestor of these species had lived more than a billion years ago. Ever since then, they and their genomes evolved separately, and a lot can happen to a genome in a billion years. Every single DNA letter can be hit by a mutation, and not just once but multiple times. Many genes can become duplicated, and many others deleted when they are no longer needed. Given how long ago these species split, and how much a genome can change over time, it is no surprise that each species would harbour many orphans. Each orphan gene could have been present in the common ancestor of these species, but then diverged beyond recognition or become deleted in one of the two species.

But this argument does not work for closely related species. Because their common ancestor lived in the recent evolutionary past, their genomes accumulated fewer point mutations or DNA deletions since they went their separate ways. Their genomes should therefore be more similar. That is indeed the case. It is another general principle of genome evolution: the more closely related two species are, the more similar are their genomes. For the same reason, one would think that more closely related species harbour fewer orphan genes, such that the closest species

– those that split only a few million years ago – might harbour no such genes.

False. The number of orphan genes does indeed decrease in closely related species, but not to zero. Even the most closely related species harbour multiple orphans. For example, once the genome of a dozen yeasts had been sequenced, including one that split from brewer's yeast less than five million years ago, brewer's yeast was still left with more than a hundred unique genes. In other species too, many orphan genes persisted as ever-more-closely-related genomes were sequenced.[16]

The fog around orphan genes began to lift with the sensational discovery that many orphan genes are indeed brand-new genes.[17] Researchers discovered this when they aligned the genomes of very closely related fruit fly species and compared them. They found many stretches of DNA that one species transcribed into RNA but multiple others did not. Many of these new transcripts also contained an open reading frame, the pattern of start and stop signals separated by multiple letter triplets that encode a protein.[18]

In addition, multiple of these newly transcribed genes diverged more slowly in their DNA text than other, non-transcribed regions of the fly genome. This is important, because it means that natural selection rejects more mutations in these new transcripts than in other genomic regions. In other words, these transcripts are important for a fly's life, even though we may have no idea what they are doing.[19] Indeed, some of these new genes are more than just important. They are essential. When researchers turned their transcription off with genetic engineering tricks, the flies lacking these transcripts died.[20]

To see where Jacob's categorical denial of *de novo* genes went astray, consider that the genome of fruit flies and many other organisms does not just contain genes, but also vast stretches of DNA that are *non-coding* – they do not encode any protein.

For example, eighty percent of the fruit fly's 200-million-letter genome does not encode protein. And a whopping 98.5 percent of our own three-billion-letter genome does not encode protein.[21] A lot of this non-coding DNA – also called junk DNA – stems from ancient pseudogenes and defective mobile DNA. It can freely accrue myriad mutations over time. These mutations erase not just the DNA's semblance to other genes, but they essentially create random strings of DNA text.[22]

Jacob did not know about the sheer amount of such junk DNA, nor did he know that its random DNA strings can be surprisingly consequential for evolution. An ingenious recent experiment illustrates what I mean. In this experiment MIT researchers used genetic engineering to attach forty different random DNA strings to an *E. coli* gene. Each of these strings was about a hundred letters long, and each had first been computer-generated and then synthesised in the laboratory. Each replaced the natural DNA that normally helps *E. coli* transcribe this gene. The researcher wanted to know whether the random string could substitute for the natural one in turning on the gene. Unlikely, one might think, because the random DNA would need to contain one of those regulatory DNA words necessary for transcription. And what are the chances of that?

Not too bad, actually. Ten percent of the random DNA stretches helped start the gene's transcription. Even more remarkable, in about half of the random DNA strings, a single additional random letter-change sufficed to turn the gene's transcription on.[23]

To understand why random DNA can be primed for transcription, it helps to know that cells produce not just one kind of regulator protein but dozens to hundreds of different kinds. Each kind of regulator protein recognises and binds not one but many different DNA words.[24] Different regulators recognise different words and regulate different genes. Thus, a mutation allowing new transcription does not need to create one specific DNA word. It only needs to create one among thousands of different words recognised by at least one regulator.

The *E. coli* experiment just mentioned proved that many stretches of random DNA can turn on transcription, and most others are just one mutation away from this ability. And these random DNA sequences were not even long. They comprised a mere hundred letters. The longer a stretch of DNA is, the greater the chances are that it already harbours such a word by chance alone, or that a single letter change would create one. In the 160 million non-coding DNA letters in a fruit fly genome, such words occur in the thousands and new mutations must constantly create new ones. The *de novo* fruit fly genes I just mentioned had experienced exactly this kind of activating mutation.[25]

All this means that newly transcribed random DNA is not inconceivable, and not even implausible. Rather, it is likely. But transcription alone does not make a protein-coding gene. What about the open reading frame? Isn't it extremely unlikely that random non-coding DNA would contain a three-letter translation start signal, followed by a three-letter stop signal, such that the two are separated by an exact multiple of three letters?

The question is easy to answer with computers, which can generate long, random sequences of the four DNA letters A, C, G, T and scan them for open reading frames. Lo and behold, open reading frames are abundant in such sequences. They are usually shorter than those of real genes, but not that short – if translated, many would encode proteins that are 60 to 150 amino acids long.[26] What is more, the DNA of real genomes is also brimming with open reading frames. In fact, it harbours many more open reading frames – potential genes – than actual genes. For example, the genome of brewer's yeast has 6,000 genes but more than 260,000 open reading frames.[27] The fruit fly *Drosophila* has 15,000 genes and more than 600,000 open reading frames. And humans have fewer than 25,000 genes but an astonishing 13.5 million open reading frames. That's one open reading frame for every 250 letters of our DNA text.[28]

Genomes are enormous workshops of non-coding DNA – millions to billions of letters. In these workshops, open reading frames are plentiful. And new transcription also evolves easily. Together, these facts show a clear path – a ten-lane highway really – to the origin of new genes. The first step is that a DNA mutation creates new binding words for transcription factors, which happens frequently in large stretches of non-coding DNA. When any one such mutation triggers new transcription, chances are that the transcribed RNA already contains an open reading frame. And voilà, a new gene has been created.

But the protein it encodes is essentially a random string of amino acids. What are the chances that this string is useful? They may be slim, but they are not nil. We know this, because some *de novo* proteins have been intensely studied. They help fruit flies metamorphose from crawling larvae into flying adults, help yeast cells repair DNA mutations, help rice plants fend off bacterial diseases and help fish survive freezing cold temperatures.[29] Unfortunately, for most other already useful *de novo* proteins, we don't know their purpose yet. That's because their discovery is too recent. We have not had decades to study them like other proteins.

Some genetic engineering experiments underscore the point that even random proteins can be useful. In one such experiment, researchers produced many peptides – very short proteins – such that each peptide comprised five different amino acids chosen at random from all twenty amino acids. They found that not just one but several of these random peptides helped *E. coli* survive an antibiotic. In another experiment, random peptides with a length of twenty amino acids helped a bacterium survive the lethal toxin nickel chloride.[30] Random peptides can indeed be useful.

Like these man-made peptides, many natural *de novo* proteins are short.[31] Such short proteins often do not fold into the orderly three-dimensional shapes known from enzymes. That too need not be a problem, however, because many proteins do just fine without this shape. In such *disordered* proteins, all or part of the amino acid string is free to flip, flap and flop around. This flexibility can even be part of a protein's job, for example in proteins that capture other, damaged proteins and help recycle their parts. The flexible amino acid chains of such garbage-collecting proteins can help trap their quarry, much like the flexible arms of an octopus can ensnare its prey.[32] More generally, we know that short proteins can perform many important tasks – they can help regulate microbial metabolism, shape plant leaves or build animal skin.[33]

While Jacob's belief that genes are never created from scratch was wrong, his intuition that gradual change is important for evolution was dead-on. It even applies to *de novo* genes, although he could not know that. New genes can be created and refined in small steps, like many other innovations of evolution (and of humans).

I already mentioned the first of these steps. These are the DNA mutations that transform an inert stretch of DNA into one that is transcribed into RNA.[34] A new RNA transcript may not encode a protein, but it can already be useful. That's because a transcript can do more than just help to make protein. Some RNA molecules help turn other genes on or off, others help defend organisms against viruses, and still others help hormones like oestrogen do their job. Most genomes transcribe thousands of RNA molecules that are useful without being translated into protein.

The next steps in a gene's birth are the mutations that create an open reading frame in those transcripts that do not already harbour one. They too will be few, because open reading frames are so abundant even in random DNA.

Most new genes are transcribed rarely, perhaps as little as once in the lifetime of a cell, but if they are useful, transcribing them more often would be better, because it would help produce more of the useful RNA or protein. Indeed, as genes are born, survive and get older over millions of years, they do become transcribed ever more frequently, thanks to mutations in a gene's regulatory DNA. Such DNA contains the words where regulator proteins bind and trigger transcription.[35] Any regulator can bind its regulatory DNA firmly or weakly. Firm binding means frequent transcription, and weak binding means rare transcription. When binding is weak, it can easily be strengthened by mutations that create stronger DNA words. One or few such mutations usually suffice to crank up a gene's transcription.

The next evolutionary steps lengthen a nascent gene. While short genes and small proteins can be useful, longer, more complex proteins are often better. They can perform their job more efficiently or flexibly – like a radio where you can adjust the volume, change the station, listen in stereo or set an alarm, as opposed to one with just an on-off button. Indeed, as new genes get older, they also tend to get longer, and all that is needed are mutations that change one of the stop signals for translation into a different three letter word.[36] From such a word, the ribosome simply continues translating until it reaches the next stop signal that the DNA happens to contain.

Each of these steps – mutations that create a new transcript, increase how often it is transcribed, create a new protein, and extend its length – is small. But together they can create a long, complex and abundant protein whose origin defies imagination unless you know that natural selection can preserve each tiny step, as long as that step improves life. Jacob didn't know – couldn't have known – how easy these steps are, because only the genome era taught us about the immensity of evolution's DNA workshops.

The origins of life's complexity, evident in millions of species from tiny bacteria to giant elephants, in complex organs like eyes and brains, and in the billions of molecules that populate each cell, boggle the mind. But understanding these origins boils down to this: we need to understand the small individual steps that lead up to each species, organ and molecule. Thanks to genomics, these steps have become obvious for new genes and proteins. Yet I write about them mainly for another reason. They embody, on this smallest scale of life's complexity, another principle at work everywhere, from the origin of mammals in a world of dinosaurs, to human inventions that come long before their time: nature creates much more than it actually uses.

Here is a clue that new genes are no exception to this rule. I already mentioned that very little DNA in our genome – about 1.5 percent – encodes proteins. I did not mention that most of the remainder – some eighty percent – is transcribed. The transcribed portion accounts for some 58,000 transcripts, three times more than protein-coding genes.[37] In other words, cells madly transcribe their DNA, far beyond their need for proteins.

Although some of these transcripts are useful, most of this molecular scribbling is pointless, because most transcripts do not benefit the organism. We know this, because mutations that alter or eliminate them cause no damage. If these transcripts mattered to life, then at least some such mutations should affect life, but they do not. Genomicists say that their DNA evolves *neutrally*. This means that life tolerates all mutations in them, and can pass these mutations on from generation to generation. At least eighty-five percent of our genome is such neutral DNA whose transcription serves no purpose.[38] Evolution is like a writer who produces millions of pages of text that nobody cares to read.

About a quarter of the transcribed DNA encodes protein, but even most of *that* DNA is useless.[39] The proteins it encodes also evolve neutrally, which tells us that they serve no life-sustaining purpose. Not all of these proteins have originated recently in evolution, but the same holds for the recent ones. For example, when researchers examined more than seven hundred new proteins in the house mouse, they found that two thirds of them evolve neutrally.[40]

If such proteins fail to prove their worth in sustaining life, then mutations will sooner or later destroy their open reading frame, their regulatory DNA, or both, just as easily as mutations created them. In fact, that happens sooner rather than later. For example, among 388 transcripts detected in a species of house mice, thirty do not exist in a closely related mouse species. And forty percent of them do not exist in the distantly related rat, which split from the mouse lineage less than twenty million years ago, a modest span of evolutionary time. Mutations destroyed many of these transcripts in the rat lineage.[41]

We also know that about half of the mouse genome is transcribed, but it's not the same half in different species of mice. Evolution turns transcription on and off very rapidly during the millions of years in which evolutionary time is measured.[42] If a part – any part – of the mouse genome is not transcribed at any one time, then within a mere million years, some mutation will have triggered its transcription. And if a part is transcribed, but that transcript is useless, its transcription won't last. As soon as new mutations damage its regulatory DNA, it will no longer be transcribed. In other words, over time – a lot of time compared to our life span, but little in evolutionary terms – nature experiments with every single tool and gadget in its gigantic workshop of non-coding DNA.

That's perhaps the biggest difference between a genome's junk DNA and the unused junk in a human workshop: junk DNA does not sit idle. Mutations relentlessly alter its letter sequence and create ever-new strings of molecular text. Some of these mutations trigger the DNA's transcription and create an opportunity for the DNA to prove its mettle in life's unending quest for the new and useful.

Imagine a genome as one of those strings of decorative Christmas lights (it would be thousands of kilometres long). Imagine also that only the transcribed parts of this genome light up, and that you could watch for millennia how individual genes blink on and off during the genome's evolution. As the millennia pass, you would see thousands of new regions blink on. They are like tinkerers' inventions, encoding new RNA or protein molecules that might be useful. But most of them are not, and they will soon flicker off again, just like most human inventions fail. Only the few useful genes stay lit, and even many of these go dark again after some time. They have become obsolete, just like inventions that were once successful become superseded by newer ones.[43]

The longer a gene has been useful, however, the greater the chances that it stays lit – survives – for a long time. A small fraction of genes survive for tens or hundreds of millions of years. These are the most important genes.[44] Some of them are central to all species. They include genes that help replicate DNA, create energy-rich molecules or build cells, just like a small fraction of human inventions – gunpowder, the combustion engine, electric lighting – have endured in their usefulness. These genes are like steady beacons that never turn off, while most lights in the enormous light show around them blink in and out of existence, some slowly, others rapidly.

Although this chapter focuses on the origin of genes, it's worth pointing out that the same pattern of evolution exists for other molecular features of life. Among them is a class of molecular switches that control what proteins do. These switches involve the small phosphate molecules we already encountered in chapter 4.

Cells can attach phosphates to proteins or remove them from proteins. And even though phosphates are tiny compared to much larger proteins, they can transform what a protein does, because they can alter its shape and activity. Phosphates turn some proteins on, while they turn others off. They activate some regulator proteins while they shut others down. They allow some proteins to assemble into larger machines, while they cause other molecular machines to fall apart.

Flipping these phosphate switches requires specialised enzymes. Different enzymes are specialised in flipping phosphate switches on different proteins. Some of them attach phosphates, whereas others cleave them off. Each such enzyme recognises specific words on a protein, sequences of letters written in the twenty-letter chemical alphabet of amino acids, which instruct it to attach phosphate to the protein, or to detach it. In other words, each word embodies a switch, and attaching or detaching phosphate amounts to flipping the switch.

Like all parts of a protein, each such switch is encoded in DNA, and mutations can change this DNA. Some mutations can destroy the switch. Other mutations can create a new one, which is easy because the amino acid words of these switches are short. They are as short as the DNA words that regulate a gene's transcription. This also means that many proteins, including random strings of amino acids, already contain such switches. And even if they do not, a single point mutation often suffices to create a new one.

The number of these phosphoswitches is staggering. Consider a humble microbe like brewer's yeast, whose genome is less than 1/100th the size of ours. Its proteins contain almost four thousand different phosphoswitches. And the evolution of these switches mirrors the evolution of new genes. Most of these switches are young – DNA mutations have created them recently. What is more, most of them are short-lived. Within a few million years they are gone again. They rapidly blink in and out of existence, like most genes do. Only thirty percent last tens of millions of years, and fewer than ten percent – the most important ones – last hundreds of millions of years.[45]

Nature incessantly produces genomic inventions like new genes or new phosphoswitches. They resemble human inventions in that very few are successful right away, many become successful only after some time, and most never become successful. DNA mutations erase them before they have proven their worth in the marketplace of Darwinian evolution.

The events that awaken some of these genomic inventions remain a mystery, just like the question why many remain dormant for a long time, or why most die before they become awakened. But we have some hints. They come from some new and useful genes whose function we know, like one that helps bacteria survive toxic nickel, or another one that helps rice plants combat bacterial infections.

They hint that a successful gene does not succeed on its own. It succeeds only at the right time and place, in the right environment, such as an environment that contains nickel or the right kinds of infectious bacteria. In other words, a changing environment may be crucial to awaken dormant genes.

This suspicion is hardened by the genome of an organism that is not just inconspicuous but barely visible, a water flea in the genus *Daphnia* so tiny that you need a microscope to examine it. In its nearly translucent body, you would see a frantically pumping heart and legs that tirelessly beat to create a current. The body's silhouette resembles a freshly hatched chick, except that its head sprouts two enormous antennae, and its rear sports a sharp thorn-like spike.

Daphnia is widespread in nutrient-rich lakes. Aquarium lovers use it as fish food, because it is undemanding and reproduces rapidly. Ecologists use it to study whether chemicals are toxic to water-living animals, because it is abundant and easy to handle.[46] But genomicists can perhaps benefit the most from *Daphnia*, because its genome has a remarkable property: more than ten thousand of its thirty thousand genes do not exist outside *Daphnia*.[47] Many of these genes originated in *Daphnia*, and then became duplicated once, twice, or more often, up to eighty times. From there, many diverged in sequence and eventually acquired new jobs.

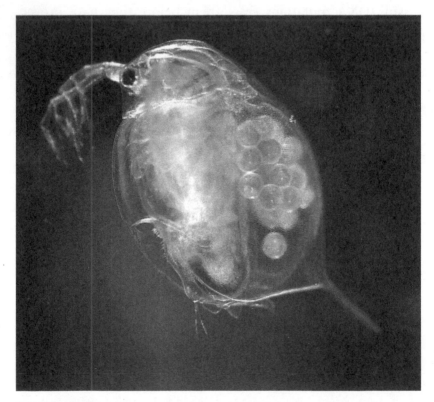

Fig. 6 *Daphnia magna*

If we want to understand what new genes do, *Daphnia*'s genome is priceless, because it contains thousands of them. To find out what they are capable of, researchers performed a complex and informative experiment. They exposed *Daphnia* to multiple environments, including environments rich in toxic heavy metals like zinc, cadmium and arsenic, environments poor in oxygen, immersed in UV radiation or swarming with *Daphnia*'s natural enemies: infectious bacteria, predatory insects or *Daphnia*-loving fish. In each environment, the researchers recorded which *Daphnia* genes were actively transcribed. A gene that is active in one environment but not in others indicates that the gene is needed there, because when organisms regulate genes, they do so according to their need. Genes active in an environment often enable or improve life in that environment. In the majority of environments, the researchers found that most active genes were new genes.

This link between new genes and changing environments is not hard to understand. The environment changes all the time. It's hot and bright during the day, cool and dark during the night. Predators and prey come and go, and so do various nutrients, toxins and countless other chemicals that arrive and disappear as they get consumed, are washed out by rain or degraded by the sun. Compared to that chaos, the inner life of a cell changes little. Not only has it already survived eons of evolution, its very purpose is to shelter life, to insulate its fragile molecules from the chaos around. A bit like an adventurous voyager who explores faraway lands rather than his placid hometown, a new gene will find many more opportunities to prove its worth in contact with the chaotic outer world, rather than within its birthplace. Even though a gene may exist only for a brief moment in evolutionary time, this moment may last millennia, and during this time, the environment changes thousands, no, millions of times. Each such change is a gift to that gene, a new opportunity to prove itself useful.

Thus, like other innovations we have already encountered, the value of a gene does not just come from some inner quality of the gene. It comes from the world into which the gene is born, a world beyond the organism's control. The beauty of genomics is that it can quantify how often new genes succeed, because genomes are so vast, and because evolution constantly creates new ones. Take our own genome. Since our own lineage split from that of chimps, evolution birthed almost eight hundred new transcripts. Many of them will die again, some of them may remain dormant for a long time, and so far, only a handful – six by one count – have become useful.[48]

We have descended in the hierarchy of nature, from charismatic organisms like ancient birds down to the submicroscopic world of genes. As we next ascend the world of culture, from the simple material culture of tool-wielding animals to the complex culture of modern civilisation, we will find the same principles at work. Humans unceasingly create new ideas, works of art and inventions, but many of them leave little trace in the historical record. If they do not become successful eventually, they are forgotten. In the world of ideas that our minds inhabit, success is rare, and dormant, dead or dying inventions are legion, just as in the molecular world of genes. And the similarities don't end there. In both worlds, today's losers occasionally become tomorrow's winners.

PART TWO

Culture

6

The crow and the pitcher

LIGHTLY ARMED WITH A TRIDENT and a dagger, the Roman glad-iators called *retiarii*, or net fighters, would have been helpless against heavily armed and shielded opponents were it not for their greater speed and that crucial net, which they threw over the enemy to entangle and topple it. They were not the first to use such a trick. Not even close. Millions of years before, nature created tropical spiders in the genus *Deinopsis* that construct an unusual rectangular net made of fuzzy spider silk. These spi-ders do not lurk passively in their net like the more familiar orb-weaving spiders. They are nocturnal hunters that dangle vertically from a thread of silk, holding the net between their legs. When an unsuspecting victim passes underneath, they cast their net on top to entangle and immobilise their prey. Their common name – gladiator spiders – is well deserved.

In the century since German psychologist Wolfgang Köhler proved that captive chimpanzees ingeniously use sticks, boxes and other tools to access food, we have learned that animals need not even have chimpanzee smarts to produce and use tools. One-inch spiders with brains smaller than a pinhead can do it, as can many other species less sophisticated than chimps. Ants pelt the nest entrance of other colonies with stones to pre-vent their members from foraging. Wasps that deposit their eggs in subterranean burrows tap a pebble against the opening

Fig. 7
A gladiator
spider

to smooth the soil and disguise the burrow. Sea urchins decorate their bodies with shell fragments, algae or rocks, not to impress the ocean fauna with their fabulous outfits, but for the opposite reason: camouflage. Octopuses take such camouflage to a new level when they crawl into the bottom half of a sunken coconut shell and cover themselves with the top half to ambush prey. Sea otters wrap crabs they catch in algae, not to prepare sushi, but to prevent their prey from scurrying off. The same animals can use a rock balanced on their belly as an anvil to crush mussels while swimming on their backs. What is more, they store these rocks in a skin pouch near their armpits that serves as a tool box. And long-tailed macaques living near a Thai Buddhist shrine floss their teeth with human hair.[1]

The use of tools – objects manipulated for a purpose – requires an ingenuity that we long believed to be exclusively human.[2] That belief is mostly self-flattery, but if it's any consolation, human tools are superior in some ways. A rock hammer is a less complex technology than a jackhammer. Also, animals power their tools only with gravity and their own energy, whereas we also harness the wind, the sun, fossil fuels and nuclear energy. Distinctions like these will persist, but many others are crumbling as we learn more about animal ingenuity. For example, humans were thought to be the only species that uses tools to make other tools. No longer: capuchin monkeys can use a rock to shape a twig to dig for insects in tree-holes.[3]

The innovations of biological evolution from previous chapters – evolutionary innovations – are as different from each other as new animal body plans and new genes, but they all emerge from a common substrate. This substrate is DNA and its change through mutation. The innovations of this and the next chapters begin with simple animal tools and culminate in the most sophisticated human creations. They are no less diverse than evolutionary innovations, and they too emerge from a common substrate. This substrate is the complex network of neurons we call a nervous system. Cultural innovations ultimately emerge from changing firing patterns of these neurons.

Nervous systems are products of biological evolution, but once created they change the rules of the innovation game, and help life break free from the strictures of biology. Just compare their speed to that of DNA mutations. The chances that any one DNA letter in a human genome mutates in any one year are below one in a billion.[4] (Only because our genomes have so many DNA letters can evolution proceed at all.) Bacteria evolve faster partly because their huge populations harbour so many hopeful inventors. But even they require days or weeks to come up with new survival tricks. In contrast, neuronal firing patterns can shift within milliseconds to bring forth a new animal behaviour or human idea.

And nervous systems can do more than turbocharge the speed of innovation. Once they have created an innovation, its spreading does not require the slow process of natural selection, which can take many generations. It can spread through a much faster process that we can already observe in the simple tool-based technologies of animals.

Important clues about this process come from species that use tools not just occasionally but habitually. For example, most chimpanzees in a population from Mahale in Tanzania forage for ants with sticks, ninety percent of sea otters in a Californian population crush snails or mussels with rock hammers, and most orangutans from Kutai in Borneo use leaves as napkins.[5] Apes and sea otters are very different animals, but what they share is a complex social life. Their individuals observe each other, and they can emulate the one individual who has figured out a new use for sticks, rocks or leaves.[6] This ability – social learning – goes a long way towards explaining how widely these species use tools.

In a 2005 experiment that illustrates the power of social learning, Andrew Whiten and collaborators designed an apparatus from which chimpanzees could retrieve a morsel of food using a stick. The scientists then chose one chimp from a group of seventeen, and trained her to use the apparatus. They did not choose just any chimp, but a female with high standing in the social hierarchy of the group. Individuals like these are the influencers of the animal world, role models for others who are more likely to follow their technological lead than to follow low-ranking juveniles, just like most people will pay more attention to the discoveries of an award-winning senior scientist than those of a first year PhD student.[7]

Once the chosen chimp had become an expert in feeding-by-stick-poking, the scientists allowed her to rejoin the group and gave all its members access to the apparatus. In a matter of days, her knowledge had spread, and most others had learned to get food from the apparatus. Social learning was essential for this success, as the scientists found out when they gave the apparatus to a second group of chimps, but without training one of them. None of the group's members discovered how to get to the food.[8]

Social learning can also create irresistible social pressure, as another experiment by the same researchers proved. They had equipped their apparatus with a second way to extract food – not by poking a stick, but by lifting a part of the apparatus with a stick. When the researchers trained a chimp from a third population in this stick-lifting technique, and then allowed this new expert back into the population, most individuals soon adopted stick-lifting. Remarkably, even though some individuals in this stick-lifting population discovered the stick-poking technique, most of them eventually abandoned their discovery and joined the group majority. In other words, these animals displayed conformity. They – like many humans – tend to mimic what most others are doing.

Social learning can even allow animals to create enduring knowledge traditions. Such traditions resemble those of human crafts like carpentry or masonry, where skills are passed from masters to apprentices over generations.[9] Consider the bearded capuchin monkeys from the Brazilian Serra da Capivara National Park. Their knowledge tradition is about the art of cracking cashew nuts and other hard foods with rock hammers. An archaeological dig at a site popular with nut-cracking monkeys unearthed ancient remnants of such rock hammers. They reveal a tradition of nut-cracking that goes back hundreds of monkey generations, or more than 2,400 years. The monkeys in this population have been pounding away at nuts since long before Columbus discovered the Americas.[10]

Social learning can even allow animals to learn from a different species, with dramatic results. Here is what one female orangutan picked up in a rehabilitation camp, where humans reacquaint animals with life in the forest after they have been orphaned by loggers or sold illegally as pets. According to one observer, this orangutan:

> hammered nails, sawed wood, sharpened ax blades, chopped wood, dug with shovels, siphoned fuel, swept porches, painted buildings, pumped water, blew blowguns, fixed blowgun darts, lit cigarettes, (almost) lit a fire, washed dishes and laundry, baled water from a dugout by rocking it side to side, put on boots, tried on glasses, combed her hair, wiped her face with Kleenex, carried parasols against the sun, and applied insect repellent to herself.[11]

With this kind of aptitude, it is not surprising that social learning is also important in wild orangutans, which use tools to harvest honey, ants or termites from tree holes. That's why orangutans that explore the forest in large foraging parties use tools more often than orangutans in smaller parties. Larger parties offer more opportunities for social learning.[12]

Social learning is important, because it allows innovations to spread much more rapidly than by natural selection. But before an innovation can spread, an animal must have discovered it. And here we face questions that we have encountered before. The first is whether tool discovery is easy or hard. Do individuals of a species repeatedly discover the same tool, or are these discoveries singular events? Did an Einstein of octopuses discover how to hide in a coconut shell? Did an Archimedes of monkeys discover nut cracking?

Some tool innovations indeed appear to be hard. For example, many chimpanzee populations in Africa crack nuts, but chimpanzees in Gabon do not. It can't be for a lack of nuts, because nuts are plentiful where they live. It also can't be for a lack of rock hammers and rock anvils, because these too are abundant.[13] Even more tellingly, in Côte d'Ivoire, among a pair of neighbouring chimp populations, one cracks nuts and the other doesn't, even though the territories of both harbour nuts and rocks. These populations are separated by the Sassandra River, which prevents animals of one population from crossing over to the other. Both of these examples hint that chimpanzees do not discover nut-cracking frequently. Instead, they discover it so rarely that some populations did not discover it at all. Perhaps that's because nut-cracking is not an easy skill. It takes chimpanzees four years of practice to learn it properly.[14]

Examples like these are reminiscent of those few unique evolutionary innovations from chapter 1 that originated only once, such as the giant moisture-harvesting leaves of the desert plant *Welwitschia*. And like such examples, nut-cracking may be an exception, because many other tool skills are easier to discover. That's evident from a study examining tool skills in nine populations of African chimpanzees, from Senegal in West Africa to Tanzania in East Africa. The study examined eleven different tool skills practiced in multiple populations, such as using a stick as a club, fishing for termites inside a termite nest, and shooing pesky flies away with a leafy branch. Each of these skills was practised in some but not all populations, and – this is important – the populations did not cover a continuous territory but formed a patchwork. In other words, each skill was widespread in at least two chimp populations separated by hundreds or thousands of kilometres, but completely absent in one or more populations in between.

How did this patchwork of tool skills arise? One possibility is that each skill was discovered only once, spread through multiple populations, but was then forgotten again in some of these populations. Given how powerful social learning is, that would be nearly impossible as long as the skill is useful.[15] More likely then, each skill was discovered multiple times, each time in a different population. If so, these skills are similar to many evolutionary innovations that we encountered earlier, innovations like latex or molars that evolution discovered more than once.

The same holds for the peculiar hunting tools of bottlenose dolphins from Australia's Shark Bay. Shark Bay is an enormous body of water, covering more than two million acres and fifteen hundred kilometres of coastline that subdivides the bay into multiple inlets. These inlets harbour deep water channels with sponge gardens that provide dolphins with their hunting tools. Dolphins use their delicate beaks to dig in the ocean floor for hidden prey, and to avoid getting their beaks scoured they first impale marine sponges with them to use as protective tools while digging.[16] Curiously, it's mostly females who use sponges in this way. DNA tests show that within a family these females form a matriline, a sequence beginning with a mother, and continuing with her daughter, grand-daughter and so on. In other words, sponge-fishing is passed on from mothers to daughters.

Sponge-fishing must have been discovered multiple times, because scientists have observed it in dolphin populations separated by more than thirty kilometres, even though female spongers rarely travel far outside their channels.[17] More important, DNA tests show that spongers form more than one matriline. Each of them was founded by a different dolphin, an innovator who discovered this new way of hunting, perhaps many generations ago.[18]

Yet another case for multiple discoveries comes from Japanese macaques on the island of Koshima, where researchers left sweet potatoes out on the beach, and one female discovered that she could wash the potatoes in fresh river water instead of just brushing off the dirt. The new skill spread rapidly through the population – a singular discovery reinforced by social learning, one might think, except for this: four other macaque troupes independently discovered potato-washing.[19] (Sensitive to the finer points of the culinary arts, many macaques also learned to dip the cleaned potatoes in salty water before eating them.)

Innovations like these also help us understand why they are easy to discover: they co-opt other skills that require little change. Japanese macaques already brushed sand off food with their hands before they discovered how to wash potatoes, and it is not a big stretch to do the same under water. Indeed, when these monkeys wash potatoes, they often hold a potato in the water with one hand and brush it with the other hand, just like they would use both hands to brush sand off it.[20]

Another kind of co-option is used by New Caledonian crows. They are famous for their ability to rip a thin strip off a hard Pandan leaf, and to use this strip as a stick tool while prospecting for insects in tree holes. This skill builds on another, less dramatic one, because the same crows can rip the leaves of Pandan trees to shreds, presumably to reach insects hiding between the tightly packed leaves.[21] From shredding a leaf it may not be far to using those shreds for a new purpose.

And then we have the ant lion, an insect larva that excavates a conical sand pit and lurks at the bottom of the pit, venomous pincers at the ready for any prey that might come its way. The pit's walls are so steep that any incautious ant stepping over the pit's rim will slide to the bottom, where the ant lion waits. And should the ant manage to scramble up the wall, the ant lion has a nifty trick in store. It flicks its head to hurl sand at the ant, which not only gets bombarded with sand grains, but slides downhill on the tiny avalanches that these sand grains trigger when they hit the ground. Devious, but perhaps not difficult, because the ant lion uses the same sand-flinging movement to dig and maintain its pit.[22]

Like many of the evolutionary innovations encountered in the previous chapters, the discovery of new tool skills cannot be too difficult when animals discover such skills repeatedly. And just as in those previous examples, the co-option of existing skills can make multiple discoveries easier. But most importantly, animals also show an innate potential for such discoveries. This potential is easiest to detect in captive animals, because they are simpler to observe than animals in the wild.

Egyptian vultures highlight this potential. They are widespread in East Africa, where they like to feast on the enormous eggs of ostriches. Some ostrich farms open to tourists demonstrate the hardness of an ostrich egg by allowing a visitor to stand on one that is half-buried. That abuse does not usually faze the egg. To impress the tourists further, a tour guide may then open the egg by more violent means, such as a power drill. Vultures are not familiar with the wonders of power drills, so when they encounter an ostrich egg in the wild, they resort to simpler tools: they throw rocks at the egg until it cracks. This rock-throwing skill is widespread among Egyptian vultures, not just in East Africa, but also as far afield as South Africa, Israel and Bulgaria. It could have spread from a single discovery via social learning, because these birds migrate far and wide. However, researchers who hand-reared two Egyptian vultures without bringing them into contact with other vultures showed that this is not likely. Once each of the youngsters had been shown that ostrich eggs harbour food, it developed the rock-throwing skill in no time and independently from the other vulture. Each youngster harboured an innate aptitude to discover this skill. And once again, the co-option of a simpler skill made this discovery easier. Egyptian vultures often crack open the eggs of smaller birds by throwing these eggs on the ground. It is not far to go from egg-throwing to rock-throwing, and the vultures' behaviour indeed hints that one emerged from the other. That's because rock-throwing vultures prefer to throw smooth, egg-shaped rocks.[23]

Better studied than the tools of Egyptian vultures are those of Galapagos woodpecker finches. They trim and smooth twigs or cactus spines before using them to pry insects from their hideouts in tree holes. In one experiment, researchers reared six woodpecker finch nestlings apart from other finches and gave them sticks to play with, as well as tree trunks with tasty beetle larvae inside. It took each of the six young birds only a few weeks to figure out how to use twigs to extract the larvae. Every single one of the birds discovered this skill on its own, all by itself.

The researchers then compared the stick-probing performance of these isolated birds with that of seven other birds that had the opportunity to learn this skill from other successful stick probers. Remarkably, the isolated birds were just as skilful. In other words, young woodpecker finches do not need to be apprenticed to others. Instead, they have a latent potential to discover how to wield a stick. They just need the right opportunity.[24]

Many captive animals display talents even more astonishing. Among them are New Caledonian crows – the same that turn leaf strips into tools. They also happen to be expert stick wielders and adroit stick-tool creators. These birds can use and trim a twig like any old Galapagos finch, but they can also craft twigs with hooks. That's no small feat, considering they do not have dexterous hands, but only a beak and legs to work with.[25] But their abilities truly shine when they face an even tougher challenge: improving a tool's effectiveness by using another tool, also known as a *metatool*.

In one metatool challenge, each of seven crows had to get a piece of meat from inside a box with a horizontal hole. The box was too deep for them to reach the meat with their beaks, so the crows needed to use a long stick to tease the meat out of the box instead. However, the stick they were given was too short to reach the meat. The problem seems insurmountable, unless you know that the wily experimenters had left a stick of just the right length behind the vertical bars of a nearby cage. Inconveniently, however, the stick was so far away that a crow could not reach it with its beak. The only solution was to poke between the bars with the short stick, move the long stick closer, retrieve the long stick from the cage, reach into the box with the long stick, and move the meat closer. An unnatural task for these birds, one that they would not face in the wild. Yet not only did all seven solve this problem, three of them did so on the first try.[26] In an even more complicated task involving three tools – a tool to retrieve a tool to retrieve yet another tool to get the food – four of seven crows succeeded.[27]

Captivity awakens a dormant potential for innovation in these crows. And not only in them. The phenomenon is so widespread that it has a name – captivity bias. Many animals, including elephants, rodents, naked mole rats, monkeys and apes, use more or different kinds of tools in captivity than in the wild.[28] Only in captivity, for example, do capuchin monkeys learn to use levers, gibbons learn to throw objects and bonobos learn to wipe fluids. I already mentioned that captive animals often live in close quarters with other animals, including humans, and can get an education by observing them. However, that cannot be the whole reason. For example, in the experiments I just described, each crow happily discovered new skills on its own, without observing others.

To understand what makes captivity special, it's useful to know that captive animals live the good life. They are fed on a schedule and don't have to constantly look over their shoulder for predators. As a result, they have plenty of energy and time – perhaps even too much – and if toys or tools are available, they will play with them and discover new uses. A life of plenty may in fact matter more than captivity itself, as a population of Japanese macaques illustrates. These macaques lived in the wild but were provisioned by researchers. And while they got their free lunches, the macaques played with stones, exactly the kind of behaviour that can help discover new tool skills. However, when the researchers stopped supplying food, and the monkeys had to fend for themselves, they gradually stopped playing.[29]

Some species that do not even use tools in the wild quickly discover their usefulness in captivity. Among them are rooks, close relatives of crows and ravens. One experiment challenged them to reach a worm floating on the surface of water inside a deep cylinder. The water level was too low to reach the worm with a beak. Next to the cylinder, however, lay a pile of small pebbles. The birds quickly learned to throw the pebbles into the cylinder, thus raising the water level until they could reach the worm.

Ingenuity like this was described more than 2,500 years ago by the ancient Greek storyteller Aesop in his fable 'the crow and the pitcher', where a thirsty crow throws rocks into a pitcher to get at the water. Aesop probably did not make this story up, because in the first century CE, the Roman naturalist Pliny the Elder described just this behaviour in a bird.[30] But today's science does more than just play catch up with millennia-old fables. It has discovered that captive rooks also easily learn other skills. They discover how to throw pebbles to release food, poke with sticks for prey, manufacture hooks to lift a basket and even employ tools to obtain other tools. The conclusion: these birds do not merely execute a hard-wired neural programme for using tools. Instead, they display a broader intelligence and a potential for discovery that is dormant in the wild, but blossoms in the right circumstances.

Captivity is one such circumstance, but even in the wild some circumstances are better than others. To see what I mean, let's return to Galapagos woodpecker finches – the birds that easily discover stick-poking on their own. In the wild, some finch populations inhabit a dry semi-desert zone near the coast, whereas others inhabit higher elevations with lush and humid cloud forests. Near the coast, stick-wielding is more widespread than in the forest populations. The reason is that the coast is arid, food is scarce and the finches need to work for a living. Not so in the forest, where food is so abundant that finches never need to labour at extracting insects from tree holes.[31]

In the same vein, sea otters use tools to crush their prey on rocks, but not all otter populations do so equally often. Sea otters from the Aleutian Islands of Alaska use prey-crushing tools much less often than sea otters that live much further south, in Monterey, California. That seems mysterious, until you learn that hard-shelled snails, clams, mussels or scallops that need crushing are more abundant in the south.[32]

A similar point can be made with capuchin monkeys that live in widely scattered populations throughout Brazil. Some populations inhabit open savannah-like grasslands with scattered trees, whereas others inhabit forests. Grassland-dwelling monkeys more often use rock hammers and anvils to crack open nuts than forest-dwellers. Part of the reason is that grassland monkeys spend much more time on the ground, which provides many more opportunities to find rocks and play with them. These opportunities are important, because like chimps, it takes a capuchin monkey years to learn nut-cracking. What is more, in some populations, hammer stones are sufficiently scarce that monkeys need to transport them to a larger anvil stone. (In these populations, monkeys can often be seen walking on two legs, lugging both a hammer stone and nuts to be crushed.[33])

All these examples teach us that diverse animals harbour a potential to innovate by using tools. They also teach us that this potential is dormant until awakened by the right environment. Can we say anything general about this environment? Does it have anything in common for animals as different as birds and monkeys?

One possibility is that food needs to be scarce in this environment. After all, folk wisdom tells us that necessity is the mother of invention, so perhaps animals innovate their way out of starvation. That's possibly true for some species, but certainly not for Brazilian capuchin monkeys. They rather need the opposite. They need plenty of opportunity to use tools. By running around in grasslands, they discover more rocks that provide this opportunity. And if necessity was the mother of their inventions, they should crack more nuts when other foods like fruit or insects are scarce. But they do not. Rather, they crack more nuts when they have more opportunities to crack nuts, when nuts are plentiful.[34]

Opportunity is also important for chimpanzees in Seringbara, Guinea. While they use tools to harvest abundant army ants, they do not use tools to feed on rarer termites. They do not even use tools for this purpose when other food is scarce.[35] Likewise, orangutans on Sumatra sometimes harvest insects from tree holes with tools, and sometimes they don't. They preferentially use tools wherever tree holes are abundant, regardless of whether they are starving or not.

In tool-using species, opportunity may be the mother of invention. The environment that awakens a dormant potential to discover new tools provides such opportunities. These opportunities can take varied forms, from abundant tool material, to abundant prey, to the safety of a zoo.[36]

All these examples illustrate multiple parallels between tool innovations and the innovations of biological evolution. Just as evolution discovered innovations like phytophagy and the hypocone multiple times, so animals can discover similar tool skills multiple times. In other words, many tool discoveries are not singular events in the history of a species, like an innovative DNA mutation that occurred only once in life's history. Instead, they reveal a species' inherent potential for innovation. This potential often remains dormant until awakened by the right environment. Life in captivity, with its surplus of time and energy, can help to create this environment. So can life in the wild, if it provides opportunities to play with tool materials. But ultimately, any tool innovation can succeed only when it proves useful. It can succeed only when abundant prey renders hunting tools worthwhile, when abundant nuts render nut-crushing tools worthwhile and so on. It can succeed only in the right place and time.

7

Numbering neurons

AMONG ALL FUNDAMENTAL HUMAN ABILITIES that underlie our civilisation, mathematics stands out. From its humble origins in Mesopotamian trade and taxation, it became the foundation of modern technology. And more than that, it helped formulate laws of nature that reach all the way down to the structure of subatomic particles, all the way up to the cosmos, and all the way back to the first days of the universe. Mathematics reveals principles that hold not just on our planet, but on any planet in our galaxy – or in any galaxy.

Little surprise then that scientists are wondering where maths comes from. And they find that maths builds on an ancient talent that has been slumbering for most of humanity's history. Like the aptitude of animals to discover new tools, this talent has been shaped in eons of Darwinian evolution. However, we know much more about its origins than we know about animal tool skills. For example, we know the neural circuits behind it. They teach us that even the most transformative innovations of our culture may have to lie in wait for thousands of years until their time arrives.

To find the source of our mathematical skills, psychologists love to study infants. The reason is that infants are still untouched by maths education. But experiments with infants pose a challenge, because an infant cannot tell you what it thinks or feels. Fortunately, a workaround exists for this problem. It uses the fact that an infant looks longer at a scene when she is surprised by unexpected events, like a ball suspended in mid-air or a puppet that suddenly disappears. Events like these violate basic laws of physics that even infants understand. And psychological experiments can build on this understanding.

In one such experiment an infant sits in front of a puppet stage. A hand enters the stage from a side-door, places a toy mouse on the stage and withdraws. Then a screen is raised on the stage to hide the mouse. After that, the hand reappears with another mouse, reaches behind the screen, places the mouse next to the first one and withdraws again, empty. Now two mice are on the stage, but both are behind the screen, hidden from the infant.

Infants may not be able to speak, but they are smart. They know to expect two mice behind the screen. But then the experimenter surreptitiously removes the second mouse through a trap door. And so, when the screen is removed, only one mouse is visible. When that happens, the infant is puzzled – it looks much longer at the stage than when two mice are visible. One mouse plus one mouse should equal two mice, so the infant knows something is off.

In an even simpler experiment, a psychologist places two small, opaque buckets next to an infant sitting on the floor. The buckets are far enough away that the infant cannot reach both at the same time. While the infant watches, the experimenter drops one graham cracker into the first bucket, and two crackers into the second bucket. Infants as young as ten months reach or crawl towards the bucket with more crackers. They know that two crackers are more than one. Infants can already do basic maths before they can speak.[1]

Experiments with infants, children and adults reveal that all of us are empowered with two number skills that are either innate or that develop in infancy. Both quantify *numerosity*, the number of objects in a set of objects. The first allows us to count small numbers of objects instantly and without being aware that we are counting. It's called subitising, from the Latin word *subitus* for sudden. Subitising works for up to three or four objects without any training.[2]

The second ability quantifies numerosity for more objects, but only approximately, hence its technical name: *approximate number system*. If you are shown an image with eighty dots and next to it another image with fifty dots, you can rapidly tell which contains more dots, even if you see the images for a fraction of a second, much too short to count every single dot. That's the approximate number system at work.

Most people can rapidly distinguish 10 from 13 objects, but they will fail to distinguish 100 from 103 objects. The accuracy of our approximate number system follows a principle known as Weber's law, named after a nineteenth-century pioneer of experimental psychology. The law applies not just to numbers, but also to our ability to distinguish the frequency of two sounds, the weight of two objects – that's where the law was first discovered – or the brightness of two images.[3] Simply put, Weber's law says that our ability to distinguish two numbers depends on the ratio between them. It's just as easy to distinguish 20 objects from 10 as it is to distinguish 200 objects from 100, because these numbers have the same ratio of 2:1. In contrast, it is harder to distinguish 120 objects from 100, because the ratio of these numbers is 1.2:1, much closer to one.[4]

All people can compare numerosities in this way, but some are better than others. The best can distinguish numbers that differ by ten percent or less, whereas others even struggle with numbers differing by twenty-five percent or more. And even though the best guesstimators are not necessarily those with the highest IQs, they also tend to be good at real maths – not just basic counting, but more sophisticated maths, the kind that manipulates mathematical symbols. For example, in one study that tested sixty adolescents for their ability to estimate numbers, the best had also been good at maths all the way back to kindergarten.[5] What is more, a small child's acuity in estimating numerosity can help predict its maths skills later in life.[6] The converse is also true. Children with dyscalculia – a learning disability specific to maths – are also worse at eyeballing numbers.[7]

Our talent for abstract maths builds on our number sense. And because this talent is uniquely human, so must be our number sense. That conclusion may seem obvious. At the very least it is plausible. Unfortunately, it is also dead wrong.

One kind of counterexample you may have heard about involves circus acts of animals that count. Perhaps the most famous one is a horse named Clever Hans, who lived in early twentieth-century Germany. Hans appeared to answer maths questions such as 'what is 12 + 12', by tapping the right number with his hoofs. He also appeared to correctly subtract, divide and multiply numbers.

The story of Clever Hans seems too good to be true, and indeed it is. One giveaway is that Hans would only get the answer right if he could watch his questioner while tapping. Another is that he would get the answer wrong if the questioner did not know the answer. Sure enough, an investigating psychologist found no maths talent but something else altogether. While Hans was tapping, questioners changed body posture or facial expression, revealing unconscious cues that reflected their mounting tensions as Hans' taps approached the right answer. Hans read these cues. When his taps hit the right answer, the tension in their faces and bodies released. Hans noticed this release as a cue to stop tapping.

Hans was remarkable not because of his maths skills, but as a reader of body language. In fact, this horse would become more influential to psychology than many a human psychologist. To this day, psychologists speak of the Clever Hans effect, and make sure to avoid it. When they work with animals and humans, they prevent their subjects from reading the experimenter's body language.[8] As a result, their experiments have become more reliable. And over time, these experiments revealed truly profound maths-related talents in animals.

Baboons live in large troupes of up to a hundred individuals that travel large distances to search for food. When they do, the troupe sometimes splits into smaller groups that merge again later. How do individuals decide whether to join one or the other group? Perhaps they follow one or the other leader, some high-ranking individual in the group? But no. A 2015 study of baboons equipped with GPS-trackers showed that individual baboons do not just follow a leader. Instead, they prefer to join the larger of two groups, perhaps because there is safety in numbers, perhaps because larger groups forage more efficiently.

But how does a baboon identify the larger group? Like a person who needs to decide in a flash, a baboon does not count large numbers. It estimates. And its chance of estimating correctly follows Weber's law: it depends on the ratio between group sizes. The more this ratio deviates from the 1:1 ratio of equality, the greater the chance that the estimate is correct. In fact, a baboon's ability to distinguish two different numbers is similar to that of a three-year-old human. In other words, baboons, like humans, have a number sense.[9]

Baboons are not the only monkeys to put a dent into human uniqueness. Macaques subitise small numbers like humans do, and they too can estimate larger numbers. And just like humans, they need not even see multiple objects at the same time to estimate their number. Seeing them one after another will do just fine. Even a sequence of sounds instead of visible objects will do the trick. What is more, macaques can also count a mixed sequence of sounds and objects.[10] Their ability is *cross-modal*. It does not depend on any one sensory modality like vision or hearing. Cross-modality is one hallmark of a true number sense.

Monkeys are not even the only animals with a number sense. American black bears can estimate the number of dots presented on a touch screen. Field mice prefer to prey on smaller groups of red ants than on larger ones. Lions estimate the size of another group of lions before attacking them. Crows choose more over fewer morsels of food. Guppies prefer to join the larger of two shoals. Male mosquitofish prefer to join groups with many females. Male frogs count how many mating calls other nearby males produce, and match their number – if a neighbour calls five times, they'll call six times to impress the girls.[11]

Perhaps the most painful discovery for the human ego: even spiders and insects with their tiny brains have a number sense. Spiders know the number of prey in their web, bumblebees register the number of flowers on a single plant, and honeybees remember the number of landmarks they pass on their way to a flower patch.[12] They even grasp the notion of zero. (So do birds and monkeys.) This is remarkable, because zero is an advanced concept in human mathematics. Some ancient civilisations did not even have a symbol for zero.[13] As the philosopher Alfred North Whitehead put it, 'No one goes out to buy zero fish.'[14]

It's easy to understand how organisms that need not count their change can benefit from a number sense. A bumblebee that counts the flowers on a plant need not waste effort to visit the same flower twice. A crow that scouts the ground for patches with many grubs will go hungry less often. A fish that joins a larger shoal is more likely to survive a predator attack. A lion that attacks a smaller group of animals is less likely to get hurt.

Like many other innovations we have encountered, the number sense has originated more than once. Humans and crows, for example, shared a common ancestor some three hundred million years ago, and since that time, their brains have evolved along separate paths. Because humans and crows process numbers in brain regions with different evolutionary origins, their number sense must also have evolved independently.[15] For the same reasons, bees can lay claim to yet another independent invention of the number sense. These multiple discoveries teach us that a number sense must be broadly valuable for survival. And more importantly, it cannot be very hard to evolve.

That conclusion is supported by another fact: a number sense does not require a complex brain. Honeybees, for example, have a mere million neurons compared to our billions. And even that may be much larger than necessary. When neurobiologists simulated the neural networks of a simple animal brain in a computer, they found that a network with as little as six hundred neurons can develop a number sense.[16]

Even more important is a 2018 observation from another computer-simulated neural network that was designed to mimic the flow of visual information through our brain. The network belonged to a class of networks important in artificial intelligence. They are called deep neural networks, because they harbour not just one but multiple layers of simulated neurons that are connected to each other. Deep neural networks excel at various tasks after having been properly trained, for example, to classify natural objects. During training, networks are shown many images of objects, and their neural connections are adjusted until they classify the images correctly.

Researchers had trained this specific network on 1.2 million images until it identified objects as different as spiders, dogs, tennis balls and necklaces correctly. Then the researchers asked a simple but ingenious question: can the trained network also count objects? It can. When presented with up to thirty different objects – dots like those psychologists use to test our number sense – different neurons responded to different numbers of dots. Some were most active for images with one dot, others for images with two dots and so on. What is more, like the brains of humans and other animals, the network's accuracy in estimating dots followed Weber's law. All this is so remarkable because the network had not been trained to count. It shows that a number sense can emerge spontaneously, apropos of nothing, for free, as a by-product of the ability to process visual images.[17] In these neural networks, the number sense is like one of those latent innovations we first encountered in the chemical networks of metabolism.

In 1919, the mathematician and philosopher Bertrand Russell marvelled at the profundity of mathematics when he wrote that 'it must have required many ages to discover that a brace of pheasants and a couple of days were both instances of the number 2: the degree of abstraction involved is far from easy.'[18]

He was wrong, but it took a century of research to discover that. As we just heard, a number sense is part of our basic innate intellectual equipment. Its development certainly did not require ages, at least not ages of human history. That's because it is not an innovation of human culture, but more ancient than humanity itself. It is not even a recent innovation of human evolution, because it occurs in monkeys and other mammals as well. Perhaps then, our unique mathematical talents, which build on this ancient number sense, are not a product of Darwinian evolution either?

This suspicion hardens when we consider how late mathematics arose in human history. Prehistoric bones dating back some forty thousand years bear regular markings that archaeologists interpret as numbers.[19] However, mathematics as we know it did not rise until much later, about five thousand years ago. That's when ancient civilisations like the Sumerians, Babylonians and

Egyptians developed sophisticated ways to write down not just words, but also numbers.[20] Humans have existed for much longer, since about two hundred thousand years ago.[21] And for most of this time, mathematics did not exist.

Even today, when teachers spend years drilling the importance of maths into the heads of Western children, an astonishing number of peoples around the world do just fine without calculus, thank you very much. A majority of 193 languages spoken by indigenous people from several continents do not know number words beyond five. Even more remarkably, among 189 aboriginal languages from Australia, more than seventy languages do not have number words beyond three – essentially they can express 'one', 'two', 'three' and 'many'.[22] Many such languages also do not know larger numbers that are composed of smaller numerals, such as the number 21, which is composed of the numerals 2 and 1.[23]

Well studied among such indigenous people are the Mundurukú of the Amazon river basin in Brazil. Not only do the Mundurukú lack number words beyond five, they do not use words beyond two for counting, but only to estimate numerosity. For example, the number word for five, which translates roughly as 'hand' or 'a handful' can represent five objects, but also six, seven, eight or nine. Because of their approximate maths, the Mundurukú cannot perform simple exact calculations like subtracting four from six.

Remarkably, even the Mundurukú do just fine when asked to judge which of two images contains more objects. On images with up to eighty objects, they do not perform much worse than Westerners.[24] Their basic number sense works like ours. It is independent of culture, education and language.

What is not independent of culture is mathematics itself. And its recent origin shows that culture, not biology, is paramount for its development. Mathematics must use brain circuits that are much older than mathematics itself. These circuits must contain a latent potential for mathematical thinking that lay dormant for tens of thousands of years, until it was awoken by the right culture.

Cognitive scientist Stanislas Dehaene calls this phenomenon neuronal recycling: cultural practices like mathematics put ancient brain circuitry to new uses.[25] Although the word 'recycling' is perhaps too strong – before maths arose, the old circuitry had not been discarded, but served other purposes – what matters is the main idea: maths exploits existing brain regions whose mathematical talents only need awakening.[26]

To find out where these brain regions are, it helps to know that thinking requires oxygen. So when neurobiologists want to visualise which brain regions are active while people think, they can monitor the flow of blood – a proxy for oxygen consumption – through the brain. One technology that allows them to do that is functional magnetic resonance imaging or fMRI.[27]

When people think about maths, several brain regions light up during fMRI. Among them is one of the many grooves that crisscross the surface of our brain. It's called the *intraparietal sulcus*. *Sulcus* is Latin for furrow, and you'd find this furrow underneath the rear half of your skull, a few centimetres down from the crown of your head, on both the right and the left side.[28]

The intraparietal sulcus is active whenever people estimate numerosity. It's active not just in adults but even in children as young as four, before they have received any maths training. In other words, it is involved in our innate number sense. And it links this sense with more advanced maths, which require us to understand the meaning of symbols like the digits '1' to '9'. We know this, because the sulcus is also active when adults compare, add, subtract or multiply digits. It does not matter whether they read numbers or hear them spoken, which highlights the cross-modal nature of mathematics.[29] Even blind mathematicians use the same brain areas to process maths.[30]

When monkeys judge numerosity, the same brain regions as in humans become active, regardless how these objects are shown – at the same time, for example, or one after another.[31] What is more, different neurons in these regions are sensitive to different numerosities. One neuron will fire intensely for five objects, while another will be most active for seven, regardless of what these objects are. Whether humans also possess such number-selective neurons was unknown for a long time, because it requires implanting electrodes into a person's brain to record individual neurons. Since 2018, however, we know from experiments with neurosurgery patients who volunteered to receive such implants that we too have numerosity-selective neurons.[32] And these neurons also become active when a person sees one of the Arabic numerals 0 to 9, provided that the person is literate and has learned the meaning of these numerals. This is remarkable, because it shows that learning can link otherwise meaningless symbols to our ancient number sense.

Monkeys can learn such associations too – within limits. In 2007, two German scientists taught monkeys to associate the symbols '1' through '4' with one through four objects, by repeatedly showing them each symbol followed by a corresponding number of dots. When the trained monkeys were then shown a symbol, neurons tuned to the corresponding number started firing.[33]

Thus, monkeys and humans use similar brain circuits, and they use them in remarkably similar ways to process numerosity, the most basic ingredient of mathematics. Professional mathematicians, in fact, who manipulate not just numbers but many other abstract symbols, use the same maths-active brain regions as children who manipulate digits, except that these regions are enlarged in mathematicians.[34]

In sum, the recent origin of maths, together with the ancient origin of its neural circuitry, show that the right environment rather than biological evolution was key to triggering this transformative innovation. This environment was a cultural one. More specifically, it was an agricultural environment, where land needs to be surveyed, books kept and taxes paid.

* * *

The innovation of mathematics also required abilities distinct from that of a number sense. Among them is language and its ability to handle symbols. Biological evolution did play a role in creating language, but what exactly that role is, and whether it is greater than that of culture, is still hotly debated.[35] We know from several sources that our talent for language is clearly distinct from that for maths. Firstly, brain imaging shows that language activates different brain regions than maths. Moreover, children with Specific Language Impairment often do just fine in maths. Conversely, children who are poor at maths often speak, read and understand language just fine. In addition, stroke patients with *acalculia*, the inability to recognise numbers or do basic maths, can often speak or understand language well. Other stroke patients have the opposite problem: their language skills are diminished, but their maths skills are intact.[36]

Although the origins of language are more mysterious than those of maths, one language-related skill is an exception. This is reading. It is yet another recent innovation where culture played a leading and biological evolution a supporting role. That's because reading emerged, like maths did, after the agricultural revolution some ten thousand years ago.[37] And just like maths, reading exploits ancient neural circuitry whose hidden talents remained dormant for a long time.

The first insights about these circuits came from nine-teenth-century stroke patients. One of them was a now-famous man known as Mr C, who was examined in 1887 by the French neurologist Joseph-Jules Déjerine. A retiree who loved to read, Mr C had read a book one morning when he suddenly found himself unable to grasp a single word. He felt he had lost his mind, because he could perfectly see the little inky marks on the page, knew they were letters, but could not name a single one. He could copy them, slowly, painstakingly, one by one, but that still did not help him to recognise the letters. Only if he traced the letters with his fingers was he able to name them.

Remarkably, Mr C could write just fine from dictation, with no more errors than others his age. Like most of us, he would check his writing during dictation, but now found that habit more than unhelpful. Because he could no longer read what he had written, it had become so distracting that he preferred to write with his eyes closed.

Just as remarkably, Mr C was disabled only in reading letters and words. He could read Arabic digits just fine, and he could perform complex calculations. He spoke fluently. His memory was intact. He could name any object he saw. When shown a sketch of one, he could identify it. Although he could no longer read a single word in the newspaper *Le Matin*, which he had of-ten read, he still recognised the newspaper by its shape.

Mr C's condition has a name. It is called *alexia*, an inability to read. (That's not the same as dyslexia, the more frequent and less severe reading impairment.) More specifically, Mr C had *pure* alexia, which does not affect other language skills, such as the ability to speak, write or name objects. Some patients with alexia can still recognise single letters, which allows them to slowly de-cipher words. Mr C was not so lucky. By the time he died four years after his first stroke – from a second stroke – he had still not recovered his reading ability.[38]

A stroke like that of Mr C is a tragedy for a human being, but a blessing for neuroscience. That's because brain injuries that affect one skill but not others can help scientists identify brain regions that are essential for that skill, and only for that skill. When an autopsy was performed on Mr C's brain, it revealed damage in a region called the *left occipito-temporal sulcus*. That name is a mouthful. It's another one of those grooves in our brain's cortex whose name comes from two major brain regions called lobes. One of them is the temporal lobe, which extends from near the temples all the way to the back of the head – the *occiput* in Latin – where it meets the occipital lobe.

More than a century after Mr C's death, we know that patients with pure alexia usually suffer damage in this region. I will call this region, for simplicity's sake, the *word reading area*.[39] Modern brain imaging reveals that its neurons are active whenever we read. Not only that, the more fluent a reader is – the more words she can read per minute – the more active the area becomes. Notably, the area does not become active when illiterate adults look at words. However, it does become active in people who learned to read as adults. That's remarkable, because it shows how easily learning can help a brain region take on a new job.[40]

The word reading area is wedged between other brain regions that help us recognise everyday objects. Specifically, it is adjacent to a region closer to the left ear whose neurons help us recognise objects related to tools. On the other side of it, closer to the head's midline, is a region that recognises faces, landscapes and objects such as houses.[41] However, the more the word reading area gets activated by words, the less it gets activated by faces, houses and tools. It's as if word reading competes with the recognition of other objects.[42]

The neurons in this region have a special power related to *invariance*, a remarkable feature of our vision. Invariance means that we can recognise objects regardless of their orientation and size. The image of a wine glass on our retina differs when we view the glass from the side, from the top, from the bottom, from nearby or from far away, but we still recognise it as a glass. Neurons near the word reading area recognise objects invariantly. They fire regardless of an object's size or orientation.[43] That's the essence of invariance.

Invariance is crucial to recognising letters and words. It is the reason why we can recognise words that vary fifty-fold in size, no matter where their image falls on our retina. Invariance is also the reason why we recognise words that are handwritten or typed, in hundreds of different fonts. And invariance is the reason why we can recognise words that are written in upper-case (GEAR) or lowercase (gear), even though many uppercase letters do not resemble their lowercase counterparts – G, for example, looks very different from g.

The word reading area lies in a region of the brain that can achieve all that for multiple kinds of objects. Its location is therefore not a coincidence, given that invariance is crucial to recognise letters. And this region is remarkable for yet another reason: it has influenced which letters and symbols humans chose for their written languages. To see why, let's take a quick look at how our brain processes images.

Any image that we see is processed in multiple and successive layers of neurons that begin in our retina. Neurons in the first layers fire in response to simple features like straight or curved lines. Later neurons respond to more complex features, such as line junctions – the places where two lines meet – and the latest neurons fire in response to entire complex objects.[44]

During this process, an object's contours are key to recognising the object. When you see a wine glass, the rim of the glass forms an ellipse that is visible from many viewing angles. The shape and orientation of this ellipse changes under different angles, but it remains an ellipse. The junction between the stem of the glass and the goblet forms a Y-like shape. And the junction between the foot and the stem resembles an inverted T. Such junctions are also visible from many angles, and they help us recognise the glass. Case in point: we still recognise a line drawing of a glass from which all lines *between* these junctions are erased. However, we fail to recognise the glass when the junctions themselves are erased.

Line junctions or intersections are everywhere around us. When you look at a flower vase while seated at a table, chances are that the vase obscures the edge of the table behind it. Where that table edge disappears behind the vase, it forms a junction with the vase, like a T laid on its side. Such junctions usually form when one object obscures another. The lines at such a junction may not be straight but curved, they could have any orientation, and they may not form a right angle, but their essential shape, their *topology*, is still that of a T.

An even simpler junction has the topology of the letter L. This shape occurs at the borders of many objects, such as in the place where two edges of a table join to form a corner. Again, from your vantage point the two lines may not meet at right angles, and they may differ in both length and orientation, but their topology is still L-like.[45]

That these junctions resemble letter shapes is not a coincidence. In fact, they helped sculpt the letters and symbols we use. This deep connection between writing and the objects around us was first revealed in two studies by cognitive scientist Mark Changizi and his collaborators. They analysed more than a hundred writing systems as different as Latin, Arabic, Hebrew and Greek. These included writing systems built on alphabets like Cyrillic, writing systems built on syllables, such as Cherokee, and logographic writing systems like Chinese, where many characters represent words. They included writing systems we use today, but also long extinct ones like Phoenician and Linear B, an ancient form of Greek. And they included both historically grown writing systems like Aramaic, and invented ones like the International Phonetic Alphabet, which was published in 1888.

In any such writing system, we can distinguish different characters by the number of continuous pen strokes needed to write them down. For example, in the uppercase English alphabet, letters like C and J require one stroke, letters like D and L require two, letters like A and B require three, and letters like M and W require four.

This number of strokes also shows a simple universal pattern, a first hint that the world's writing systems are no random accidents of history. Specifically, the number of strokes per character varies little among characters and writing systems. Most characters can be written with three strokes, some with one or two, but none with more than four, regardless of the writing system.

I already mentioned that natural objects and their junctions have a topology, an essential shape. Written symbols too have an essential shape that does not change with the length, angle or orientation of the pen strokes that create them. For example, two strokes can only form a T, an L or an X. The strokes need not be straight lines and their length may vary, as may their angles and the orientation – the T could stand on its head, the L could be toppled to the side – but disregarding such variations, two strokes will form these and only these three shapes. From this point of view, an L is topologically identical to a Greek Γ, and a T is topologically identical to an Armenian Ⴙ.

Symbols that need three strokes also come in limited shapes, but their number is a bit larger. To be precise, three strokes allow for thirty-two topologies that include the Latin F, the Greek Δ and the Cyrillic Л.

When Changizi analysed how frequent different shapes are in different writing systems, he found that some shapes occur much more frequently than others. The shapes L and T, for example, are frequent. Remarkably, they are not just frequent in alphabetic and syllabic writing systems, but also in very different writing systems like Chinese. In contrast, other shapes like that of the Greek Δ are much rarer, and this also holds across different writing systems. Shape frequency follows a universal pattern.[46]

This universality is astounding, but even more astounding is another universal pattern. It emerged when Changizi compared written symbols with natural objects. To this end, he studied dozens of images showing natural scenes, including many from humanity's ancestral homeland, the African savannah. Changizi analysed the different shapes in these images that emerge when different lines – usually edges of objects – meet or cross. He found that the shapes that are frequent in these natural scenes are also frequent in writing systems. For example, the L shape is not only the most frequent in writing systems but also in nature. The T shape is the second-most frequent in writing and also in nature. Much rarer is the Greek Δ, both in writing and in nature, because few object boundaries create lines meeting in a perfect triangle.[47]

To find out whether patterns like these could be a coincidence, Changizi also created computer-generated drawings of random lines criss-crossing a screen – they resemble a bundle of twigs thrown on the ground. Changizi then examined the geometric shapes that these lines form. He found that the frequent shapes found in these drawings are very different from those in natural images and from those in writing systems. Likewise, when Changizi studied the shapes of scribblings produced by toddlers, he found that their shapes are very different from those in natural images and written characters.

Changizi's work connects the natural world to written language via essential shapes. Such shapes help us recognise natural objects regardless of their size and orientation. And because these shapes are so crucial to recognise objects, they also found their way into our written symbols.[48]

Once you know this, it becomes clear why the same brain region is involved when child after child after child learns to read, even though writing and reading are no direct products of Darwinian evolution. It also becomes clear why the same region is involved regardless of whether this child learns to read English, Arabic or Chinese. The neurons in this region are perfectly suited to be reused for reading no matter in which language the child reads.[49] In the words of Stanislas Dehaene, 'our cortex did not specifically evolve for writing – there was neither the time nor sufficient evolutionary pressure for this to occur. On the contrary, writing evolved to fit the cortex.' And 'we did not invent most of our letter shapes: they lay dormant in our brains for millions of years, and were merely rediscovered when our species invented writing and the alphabet.'[50]

So here is yet another sleeping beauty, a dormant talent like that for mathematics, just waiting for a cultural environment in which it becomes useful. Unlike the latent talents of animals to use tools, the awakening of these talents was transformative. It permitted human culture in all of today's complexity, with millions of recorded creative works, from poems and novels to theorems, scientific discoveries and patents for technological innovations that are transforming the planet.

Writing and maths are only two well-studied among countless human innovations whose explosive success has been driven by culture rather than biology. Playing the piano, riding a bicycle, dancing the ballet, piloting a plane, painting a landscape, driving a car, playing a video game, composing a melody, weaving a carpet, wiring a circuit board, interpreting an X-ray image, programming a computer, blowing a glass vase, designing a skyscraper, building a precision watch: all these activities have no precedent in our primate evolutionary history. Whatever skills they need must use pre-existing brain circuitry, circuitry that normally serves other, more mundane purposes, but with a hidden potential to be used for a purpose that the world had never seen. And this potential remains dormant until its time has come, when a cultural innovation needs it.

If the seemingly endless versatility of our primate brain is still hard to grasp, perhaps an analogy with another versatile organ – the human hand – will help. The hand is a tool with forty-eight moving parts – twenty-seven bones and twenty-one muscles. It is different from a chimp's hand, but the differences are subtle rather than radical.[51] One of them is our ability to rotate the thumb until it opposes the palm, which allows our hand to wrap itself around an object. Another one is that our thumb is long relative to the other fingers, which allows the pincer-like precision grip between the tip of thumb and the tip of any other finger. Yet another is that our distal phalanges – the bones at the tip of each finger – are wider, which makes gripping easier.[52]

These and other subtle features gave our hand the dexterity that surgeons need to remove a tumour, that metalworkers need to operate a lathe, that watchmakers need to assemble intricate time pieces, that carpenters need to create complex cabinets, and that virtuosos need to play the violin. These and countless other skills were not preordained by our primate history. They emerged with human culture, even though they rely on an organ that predated this culture. If a tool with fewer than fifty parts can be this versatile, what about a brain with nearly a hundred billion neurons? What other skills lie dormant within, skills we have not even dreamed of?

8

Concealed relationships

WHEN ALBERT EINSTEIN CALLED QUANTUM theory 'the highest form of musicality in the sphere of thought', he was not alone in linking quantum theory to music.[1] Before him, others had discovered a deep connection between the two, with profound consequences for physics. Erwin Schrödinger, Louis de Broglie and other fathers of quantum theory used a musical analogy to help explain the properties of atoms and electrons. In this analogy, the smallest constituents of matter behave *as if* they were vibrating strings. Atoms only emit light at specific 'quantised' energies, just like a string vibrates only at specific frequencies. Atoms are stable, because the electrons that surround the nucleus behave like standing waves. Atoms blasted with electromagnetic energy resonate at specific wavelengths, a bit like when you pluck one string in a stringed instrument and others start to vibrate as well.[2]

A scientific analogy like this is far more than a figure of speech. It is an important discovery in its own right that can help formulate fundamental laws of nature. Analogy and its close cousin in literature – metaphor – emerge from an abstract thought process. This process – as fundamental to the human mind as our talent for maths – is the subject of this chapter.

The sleeping beauties of the preceding chapter were the neural circuits that power our mathematical and many other talents. They took on new roles as our civilisation and culture became ever more complex. The sleeping beauties of this chapter are more abstract. They are hidden relationships among objects. Such relationships lie dormant until we discover an analogy or metaphor that reveals them to us. Because analogies and metaphors permeate our language, such relationships are everywhere. Human creators, from poets to scientists, ceaselessly discover new ones, and our mind's talent to awaken them is nearly bottomless. Recent experiments even hint where that talent comes from. It's another example of a brain's ability to co-opt ancient neural circuits, but different ones from those that power maths.

Successful analogies go all the way back to antiquity. The Roman engineer Vitruvius already compared sound that spreads when an orator speaks on a stage to the water waves that spread when a rock hits water. Sound waves get deflected at obstacles, just like water waves do, and their deflection can degrade sound quality through echoes and other kinds of interference. The analogy had practical value, because it helped Vitruvius improve theatre acoustics.[3]

When the seventeenth-century English physician William Harvey developed the idea that blood circulates throughout our body, he compared the heart to a pump – a heretic idea at the time. When the nineteenth-century French scientist Sadi Carnot imagined an engine that transforms heat into motion, he argued that heat behaves like a liquid, flowing from a hot to a cool object like water flows down a waterfall. And when Charles Darwin developed his theory of natural selection, he leaned on the analogy of breeders that improve crops and cattle.

Powerful analogies do not describe superficial resemblances – *a head bald as an egg, lips red as rubies* – but deeper relationships between objects, even though the objects themselves may be very different. An orator brings forth sound waves like a rock brings forth water waves, even though an orator does not resemble a rock in other ways. The walls of a theatre reflect sound, just like a sea wall reflects water waves. Reflected sound waves travel back to their source, just like water waves do. Obstacles in a theatre create interference and degrade sound quality, just like ships and other obstacles render water waves choppy.

In the technical language of psycholinguistics – a hybrid research field between psychology and linguistics – analogies map one relationship onto another. The first relationship is that between concepts like waves and water in what is called a *source domain*. The second is a relationship between concepts like sound and air in a *target domain*. The nature of these concepts is not important – water is different from air, strings from atoms. What is crucial is the *relationship* between them in the two domains.

The most powerful analogies map not just one but multiple relationships from source to target. These are the analogies that enable discoveries in the target domain, like the strings that helped physics explain the light emissions of atoms. To identify such analogies is the highest of arts, because analogies can also easily mislead. Alchemists, for example, compared the sun to gold, because both are yellow. They also compared Saturn to lead, because it moves slowly through the sky, like a heavy, leaden object might. Such superficial analogies stifled the development of modern chemistry. They had to be abandoned before modern chemistry could hatch from its alchemical cocoon.[4]

The most useful analogies also rely on the old, familiar and concrete to help us understand the new, unknown and abstract. Some scientific analogies explain new phenomena like atomic spectra in terms of older physics, like the mechanics of oscillators. Others take us straight back to actions as primal as throwing a stone to explain gravity, or to phenomena as elementary as falling water to understand heat.[5]

Metaphors do for literature what analogies do for science, and they have the same goal. A metaphor like *her scalding insult* is a more compact expression than the analogy *her insult felt like scalding water*, but both express a relationship between concepts in a source domain (here: language) and a target domain (here: fluids). This goal is even encapsulated in the meaning of the word metaphor – to carry something from one place to another.

Just like analogies, metaphors are more than just a writer's tool to delight a reader. They are a – perhaps *the* – foundation of abstract thought. That's what linguists like George Lakoff discovered when they identified and analysed thousands of metaphors in everyday language. Many metaphors help our minds grasp abstract ideas by mapping them onto concrete concepts. Such metaphors are also called *conceptual metaphors*. They permeate our language. A typical spoken text of a thousand words, for example, may use about a hundred metaphors.[6] That's one metaphor for every ten spoken words. We use most of these metaphors like we breathe air – unthinkingly.

Linguists classify metaphors, and these classifications underscore how pervasive metaphors are. They are best illustrated with examples, many of them from the work of George Lakoff and psycholinguist Steven Pinker.[7]

The metaphors *scalding insults, words flowing from a pen* and *showering somebody with praise* all have a common theme. First, they share the same source domain. That is, concepts like *insults*, *words* and *praise* belong to the domain of language. Second, they share the same target domain. This is the domain of fluids – a fluid can be scalding, can flow and can be used to take a shower. In other words, all these metaphors fit the same template, which can be encapsulated in a prototype that links language to fluids. Linguists often write this prototype in capital letters: LANGUAGE IS A FLUID.[8]

This prototype is only one among hundreds we use to construct metaphors. Others include IDEAS ARE FOOD (*the meaty part of the book*), THE MIND IS A CONTAINER (*keep it in the back of your mind*), THINKING IS MOVING (*we are approaching the solution*), MORE IS UP (*stock prices are rising*) and GOALS ARE DESTINATIONS (*victory is near*).[9] Many other such prototypes are known, each with countless metaphors following the prototype. Their sheer number highlights that our minds relentlessly detect similar relationships among very different kinds of objects. But why? Is that a lesson about our minds or about the world around us? Perhaps both. Metaphors like LANGUAGE IS A FLUID illustrate our minds' ability to create pleasing new relationships, and this very ability may help us discover the complex and useful analogies of science, where sound waves behave like water waves, and changing heat behaves like falling water.

Among all these metaphors and prototypes, the most fundamental ones stand out. One of them revolves around an emotion that is fundamental to our social lives: love. Its prototype LOVE IS A JOURNEY is embodied in expressions like *our relationship has hit a dead end*.[10] In this metaphor, the lovers are travellers and the relationship is their vehicle. The metaphor maps the common life goals of the lovers to a common destination (*we have come far together*), and also maps difficulties in the relationship to slowdowns (*we're spinning our wheels*). During the journey a crossroads may be reached (*we went our separate ways*), or the vehicle may be abandoned when the relationship ends (*I am bailing out*).

LOVE IS A JOURNEY is also called a *generative* metaphor, because one can easily generate new, unconventional instances, such as *Mary left the sinking ship because John had not proposed after three years*. In skilful writing, such a new instance creates the spark of delight that we know from great poetry. However, if a new metaphor becomes widely used, its novelty fades. Eventually, it may become so conventional that we no longer realise its metaphorical nature. Such metaphors (*we're stuck*) are the living fossils of language evolution. They litter our sentences in the thousands.[11]

Other fundamental metaphors go far beyond human life. They involve the most basic concepts that help us make sense of the world. One of them is causation, a concept so fundamental that the philosopher David Hume called it 'the cement of the universe'.[12] Causation is not just fundamental, but also abstract, and metaphors can render it more concrete. They map causation onto a concept closer to our experience. This is the concept of force, like the force we exert when we kick a ball, or the force that we experience when the wind blows. The prototype for such metaphors is CAUSES ARE FORCES. Examples include *the ball broke the window* and *the cop helped the child across the street*.[13]

Another profound class of metaphors links time to space. Compare the sentence *they moved the fridge forward one metre* to the sentence *they moved the wedding forward one week*. The latter sentence refers to a wedding as if it were a physical object. The metaphor's prototype: TIME IS SPACE. Some metaphors in this category treat events in time as if they were moving objects, as in *Christmas is coming up on us*. Others refer to events in time as if they were objects in a landscape through which we move, as in *we're coming up on Christmas*.[14]

TIME IS SPACE is one of those all-pervasive metaphors that we use unthinkingly. We don't think of *the phone call was short* or *the lecture was long* as metaphors, but they are. Even a two-letter preposition can embody the metaphor, as in *the meeting was at two o'clock*, where the 'at' refers to two o'clock as if it were a location in space. TIME IS SPACE works well as a metaphor because time is more abstract than space. We live in the present, but have to remember the past and imagine the future. In contrast, we perceive space more directly. We know our location in it and experience our movement in it. Not surprisingly then, people express time in terms of space much more often than the other way around.[15]

Metaphors like TIME IS SPACE exist in language, but their roots are deeper than language. They reflect how we perceive the world. We know this from psychological experiments that study the unequal relationship – time is space, but space is not time – I just mentioned.

In one such experiment, participants were asked to observe a horizontal line of pixels that appears on a computer screen, grows from left to right and becomes longer before disappearing again. A new line appears, grows and disappears. Each time a new line appears, it grows at a different speed and to a different length. If it grows slowly but for a long time, it becomes long before disappearing again. If it grows slowly but for a shorter time, it may grow only to intermediate length before disappearing. If it grows very rapidly but only for a brief amount of time, it can also grow long, and so on.

In the experiment, after a participant saw a line appear, grow and disappear again, she had to either estimate by how much the line had grown, or the length of time the line had been visible. The catch: she was not allowed to express this estimate in language. For example, to estimate by how much the line had grown, she had to use a computer mouse to move a cursor across the screen, covering the length of the line before it disappeared.

The resulting estimates were imperfect, but precisely these imperfections can help us better understand how our minds link time and space. Whenever a line had grown long, participants estimated that it had been visible for a long time. However, that need not be true, because a fast growing line might be visible briefly but still grow long. Conversely, if a line had not grown long, the participants estimated that it had been visible only briefly. That too need not be true. In other words, their minds substituted space – length grown – for the time a line spent growing on-screen.

Remarkably, the opposite was not true: how long a line was visible did not influence its estimated length. In other words, the participants' estimates of space and time were asymmetrical, just like in language. They estimated time in terms of space, but not so much the other way around.[16] This parallel hints that metaphors and analogies can teach us about more than language. They may reflect some basic facts about the physical world, or at least about how our brains perceive this world. Indeed, this possibility is confirmed by fascinating neurobiological experiments.

Our mental abilities must ultimately be reflected in the firing patterns of our neurons, but sadly, we know little about the neurobiology of metaphors and analogies. For example, we don't know how neuronal circuits transfer concepts between different domains of experience, one ability that is central in creating metaphors and analogies. However, some intriguing experiments hint that this talent relies on circuits that are no less ancient than those required for maths and reading. They are the circuits that help us navigate the world.

To understand these experiments, one must first know that we humans rely on two specialised kinds of neurons to navigate the world, and that we share both kinds with other mammals. The first kind is located in a brain region called the hippocampus. Its neurons are called *place cells*, and their name says it all. A place cell fires when an animal finds itself in a specific place. Different place cells fire in different places.

The second kind occurs in a region called the entorhinal cortex. Its neurons are called *grid cells*, and their firing pattern is more peculiar. When a rat explores an area by walking on the ground, any one grid cell behaves as if the area was subdivided into hexagons, and whenever the rat passes a corner – any corner

– of one of these hexagons, the grid cell fires. It's as if each grid cell encoded a hexagonal tiling of the rat's environment. There are many ways to tile a surface with hexagons, and different grid cells encode different tilings – they differ in the locations of the hexagons' corners. Any one grid cell fires when the rat is located on one of the corners of 'its' tiles.[17]

Together, place and grid cells help rats, humans and other mammals navigate the space around them. But that's not their only talent. Like those moonlighting enzymes we encountered earlier, they can also participate in other tasks, abstract ones unrelated to their main job. This became evident in a psychological experiment where volunteers were shown computer-generated cartoon images of a bird. The volunteers were then trained to use a computer program that allowed them to manipulate two aspects of this bird's body shape: the length of the neck and the length of its legs. Through this manipulation, they could morph the bird into various shapes.

Once the participants had been trained, they were shown a video in which the bird's body changed shape. At the same time, their brain activity was recorded. While the participants were watching the video, their grid cells became active, as if the participants were walking through space, even though they did not move. To explain this puzzling pattern, the researchers analysed the brain recordings. They found that the participants' brains had discovered – or created? – a space that was different from the physical space we move in. It was an abstract space, a space of two concepts: leg length and neck length. Analogous to the two north-south and east-west axes of the space around us, this space had two dimensions that corresponded to the two dimensions neck length and leg length, the very quantities that the participants could manipulate. The participants had not been told beforehand about such a conceptual space, much less that it has two dimensions. Rather, their brains had discovered this space automatically, treating its conceptual dimensions as if they were spatial dimensions. And they navigated this conceptual space with the same neural equipment that mammals have used to navigate the world for millions of years.[18]

This ability to create an abstract conceptual space is not uniquely human. Other animals have it too. They can learn to use old circuitry for a new skill that is superficially different but deeply similar to an old one. An experiment with rats proves this. In this experiment, scientists trained a rat in a skill that would be utterly useless in the wild, just as useless as manipulating cartoon birds is to humans. To this end, they designed an apparatus that played a sound when the rat depressed a lever. If the rat released that lever, the sound would stop. If the rat kept the lever depressed the sound would not only continue, its frequency would also increase over time. The longer the rat kept the lever depressed, the higher the sound's pitch would rise. The rat's useless task: release the lever only once the pitch had reached a certain frequency. If the rat succeeded at this task, it was rewarded with a sip of water.[19]

The experimenters wanted to identify the neurons that are necessary to accomplish this task. Lo and behold, these were the same grid and place cells that we just heard about. While the rats performed their task, listening to the sound whose pitch increased, specific grid and place cells fired. Which of them fired changed as the sound's frequency increased. Some fired at lower frequencies, others at higher ones. And the same neurons that had learned to respond to these frequencies were active when the rats were allowed to walk around and search for food, helping the rats navigate space.

The rats' brains had mapped a gradual change in frequency to a change in the firing patterns of neurons needed to navigate space. In other words, they had discovered the neural equivalent of an analogy, between sound that changes along the single dimension of frequency and a body that changes its location along one dimension in space. They had discovered that a change in space can be a useful proxy for a change in sound frequency.

These experiments show that the brains of rats and humans can readily use neurons with one job to do another job. The tasks were superficially unrelated, but deep down they were linked by the same kind of similarity that enables TIME IS SPACE and hundreds of other conceptual metaphors, similarities like that between the space around us and a two-dimensional conceptual space. That the needed skills can be learned easily, even though they would be useless in the wild, hints that this ability is not an adaptation for survival. Rather it results from the mammalian brain's ability to transfer relationships between different domains of perception. Sadly, we know little about the evolutionary origin of this ability.

These were just two experiments and two tasks, but they come from a deep well of dormant potential, as demonstrated by a large body of human brain imaging data. This data stems from more than a thousand brain imaging experiments that measured the activity of more than a hundred different regions of the human brain while people performed different tasks. It shows that any one region is on average involved in more than eight tasks, which can be as different as seeing, hearing, remembering and reasoning.[20] In other words, the brain's neural real estate is not neatly subdivided into parcels dedicated to a specific task, like the swappable circuit boards in a computer. Different parcels work on overlapping tasks, and for this reason, any one region may help solve multiple tasks that are superficially unrelated but deeply similar.

The circuits in these experiments underscore the power of neural networks: the latent abilities of old circuitry can serve surprising new purposes. But that's not the main reason I mention them here. They also provide a glimpse of how our brain can reveal new relationships between different domains like space and sound, as when it discovers new metaphors and analogies. These relationships remain hidden, inaccessible to us, until a brain circuit has revealed them. The consequences of their discovery can be as ephemeral as a pleasurable turn of phrase or as profound as a new law of nature.

The need to map new concepts onto old ones can arise in many circumstances. It may arise in a psychological experiment, a college science problem or a quest to make sense of the world. In all these situations, analogies can help us make sense of the unknown. These situations may have little in common except for one central theme: the world around us, in the form of a task we face, is the crucial stimulus to awaken a previously dormant relationship.

9

Reinventing the wheel

AMONG THE COUNTLESS INNOVATIONS THAT humanity has brought forth, the ones that genuinely transform human life stand in a league of their own. And no, I am not thinking about recent disruptive innovations like the transistor and the internet, but about ancient innovations like the hand-axe and the wheel. Like their more recent counterparts, they must have spread instantly, and quickly transformed the societies they touched.

Or so one would think. It's certainly plausible. But the plausible can deceive. Just consider the history of wheeled transport.[1]

Wheels first appeared independently in the Middle East and Eastern Europe some 3,500 years BCE, but their ascent was neither swift nor universal. The ancient Egyptians, for example, knew about wheels, because they used potters' wheels and traded with Mesopotamia, where oxcarts and battle chariots had been used since before 3,000 BCE. However, it took them a thousand years to adopt wheeled transport.

Or take the Olmecs, an ancient civilisation that blossomed in southern Mexico until 400 BCE. Like the Egyptians, the Olmecs were familiar with the wheel, even though they did not cart their wares around. We know this because their craftsmen created many wheel-bearing artefacts, but these artefacts were not carts, wagons or wheelbarrows. They were small animal figurines, ceramic effigies of animals like dogs or cats or monkeys, with wheels attached to their legs. They look like toys, and they probably were, but they could also have been religious symbols or ritual objects – we are not sure.

As for the reasons, we are not sure either. Some historians argue that the Olmecs had no beasts of burden strong enough to pull a wheeled cart. But that neglects the human beast of burden, which had been pushing wheelbarrows in China since at least 1,200 CE. Others have argued that it was the lack of a paved road network on which wheels could shine. But Europeans used carts long before the Romans began to pave their empire around 300 BCE. Perhaps the truth is simply that pack animals and human porters were cheaper and less tiresome than broken axles, overturned carts, deep ruts, potholes and other nuisances of life with carts and chariots.

Even where wheeled transport had gained a foothold, it could still plummet back into oblivion. That happened in its very birthplace, the Middle East, where wheeled transport had all but gone extinct by the first century CE. In the Sahara, where oxcarts had been widespread in earlier centuries, they fell out of use in the first millennium CE.[2] And in China, which had adopted war chariots from Eurasian peoples by 1,200 BCE, the chariot had disappeared again by 1,200 CE. Some of that cultural forgetting had good reasons. Cavalry turned out to be more effective than horse chariots, and camels were more effective than oxcarts in the Sahara.

All this goes to show that the wheel was not a technological slam dunk. Arguably, it blossomed fully only when the railroad and the automobile industry rose in the nineteenth century. Only then did rail and road networks begin to transform the world's landscapes, several millennia after the first discovery of the wheel.

The invention of the wheel may have required the abstract thinking of the previous chapter, or it may have been one of countless accidental discoveries in the history of technology. We may never know. However, we do know that the wheel, like simpler animal technologies, differs from the innovations of biological evolution in some ways – it did not originate in DNA but in somebody's brain, and it spread not just by natural selection but by social learning. These differences are important, but in the eyes of many researchers, going back all the way to Darwin's time, they are less important than the similarities between evolutionary and cultural innovations. In their view, evolutionary innovation and human innovation result from a Darwinian process, where variation – new kinds of organisms or new ideas – arises more or less randomly, and useful variants are kept, either by unthinking natural selection or by a judicious mind.

The evidence for this view has been mounting for decades, coming from psychologists, historians and bibliometricians who study patterns of failure and success in scientific publications. Here is not the place to make a detailed case for Darwinian creativity, because that case has been made elsewhere, for example in my previous book *Life Finds a Way*.[3] But it is the place to highlight *one* similarity between technological and evolutionary innovation, because it is linked to the universality of sleeping beauties: multiple discoveries.

We heard that innovations come easily to evolution. They come just as easily to culture, and multiple discoveries are exhibit A for this claim. The wheel, discovered in the new and in the old world, is only one among hundreds of examples. Even more ancient is agriculture. It has at least eleven independent origins, including in the Middle East, China, Africa and New Guinea, with crop plants as different as wheat, rice and corn. (As we heard before, the innovation of agriculture also has parallels in the natural world, because ants and termites too discovered it during their biological evolution – they both keep elaborate fungus gardens – and they did so independently from us and from each other.[4])

We also heard earlier that evolution discovered latex multiple times in different species. This multiplicity too has a parallel in human history: the commercially useful form of latex, i.e. natural rubber.

The milky white latex bleeding from wounded rubber trees is chemically unstable. To create useful (because stable) natural rubber from it requires a process called vulcanisation. When the inventor Charles Goodyear accidentally discovered this process in 1839, he triggered a commercial revolution of natural rubber products. They became so crucial to the industrial revolution that, a century later, the president of the Goodyear Tire & Rubber Company could call rubber the 'flexing muscles and sinews' of industrial society.[5] By the time of World War II, a tank needed eight hundred pounds of rubber, and a battleship hundreds of tons. Rubber had become the single most valuable commodity imported into the US.

A neat story to show how the explosive success of rubber had to wait only for a singular discovery. Unfortunately, that discovery was not quite so singular. Pre-Columbian Amerindians had beaten Charles Goodyear by almost 3,500 years. They had discovered how to heat raw latex with the juice of a morning glory vine to achieve vulcanisation. The rubber products they produced included figurines, rubber bands and rubber balls for ritualised ball games that enacted their creation myths.[6] But because their innovation came too early, its impact on humanity was negligible.

Examples of multiple discoveries become ever more plentiful as we approach the present, perhaps because our historical records become better, perhaps because the pace of innovation is accelerating. The pendulum clock was invented at least three different times, the thermometer seven times, the telegraph four times and radar six times.[7] Examples like these are indelibly linked to the twentieth-century sociologist Robert K. Merton, who was fascinated by multiple discoveries. That's why they are also called 'Merton's multiples'. Curiously, he was not the first to discover the importance of multiple discoveries, nor did he claim to be. Merton himself found eighteen previous reports that highlight how frequent multiple discoveries are. One of them was a 1922 article that listed no fewer than 148 examples. Merton's multiples were themselves discovered multiple times.[8]

Merton argued that multiples are so frequent that singletons – unique discoveries – are the exceptions that need explaining. Indeed, careful historical research often reveals that apparent singletons really are multiples. Such research also shows that some inventions are unique only because an inventor's competitors learned about a parallel discovery, became discouraged, and dropped out before publicising their own version.

Many individual inventors would object to the general point here that innovations come easy. After all, innovating seems difficult, even for some of humanity's most prolific inventors. Thomas Edison famously had to try six thousand different materials before stumbling upon bamboo as a stable filament for incandescent light bulbs.[9] And here is how John Backus, inventor of the Fortran computer programming language, describes his creative process: 'You need the willingness to fail all the time. You have to generate many ideas and then you have to work very hard only to discover that they don't work. And you keep doing that over and over until you find one that does work.'[10]

But what feels arduous for individual innovators may look very different from a higher vantage point, like that of a historian studying an entire historical epoch. From that vantage point, Merton's multiples show that most inventions and discoveries are not hard-won singular events of history, but almost inevitable products of their time.

Among almost two hundred instances of his multiples, Merton also found about a third where more than ten years elapsed between two independent discoveries. That's a hint that early discoveries are often ignored or forgotten – sleeping beauties. The obvious question is why. Fortunately, this question is easier to answer with a detailed historical record rather than an incomplete fossil record.

Among such examples is the cure for scurvy, a deadly disease that an eighteenth-century observer called 'the plague of the sea', because it killed countless sailors who spent month after month on the open ocean.[11] Its symptoms start innocuously enough with a lack of energy and gum pain. They crescendo into muscle pains and loosening teeth. And they climax in a menagerie of ailments – festering wounds, bleeding, convulsions – that will reliably end in death.

Scurvy's death toll became most evident in the age of detailed naval records, and most dramatic when colonial powers began to explore the oceans in search of new territories, farther and farther from home, their ships spending more and more time away from land. When circumnavigating the world in 1520, the explorer Magellan lost 208 of 230 men. Two centuries later, in 1740, on an expedition to harass the Spanish in the Pacific, the British naval commander George Anson lost 1,300 of almost 2,000 men. In both expeditions, scurvy was to blame for most deaths.[12]

A potent cure for scurvy – citrus fruit and fresh vegetables – was not just discovered and forgotten, but rediscovered and reforgotten more than once. Vasco da Gama discovered it in 1497 – together with a sea route to India – in the form of oranges, but his knowledge did not spread.[13] In 1614 a British naval handbook entitled *The Surgeon's Mate* endorsed it too, but its words must have fallen on deaf ears, because even as late as the eighteenth century, the British navy lost more sailors to disease than to battles, and the major culprit was still scurvy.

A third discovery came in the form of a decisive medical experiment, a milestone in the history of medicine. Sadly, this one made no difference either. The experiment was the first controlled clinical trial, a trial of potential remedies for scurvy

Fig. 8 James Lind

performed by naval surgeon James Lind in 1746.[14] At that time, Lind served on a British warship where 80 of 350 men found themselves bed-ridden – really, hammock-ridden – from scurvy. He isolated twelve sailors that were dying from scurvy, and fed each of them a basic diet supplemented with one of six different antiscorbutics – potential remedies for scurvy. Two sailors received oranges and lemons, another two vinegar, another two cider, a fourth pair drank seawater and so on. After two weeks, the first pair had almost completely recovered. Among the other pairs, only those who had received cider had improved a bit. A clear-cut result: citrus fruit can cure scurvy.

Lind summarised his research in his 1753 *Treatise of the Scurvy*. He was well-connected to the Admiralty, which could influence the fleet's health policy, but even so the Admiralty ignored the connection between citrus fruit and scurvy for another half a century.[15] Some historians claim that Lind was undermined by competitors who favoured other antiscorbutics. Others saw Lind as too far ahead of his time, an intellectual revolutionary held back by countless reactionaries who resented his genius. The truth is more complicated.

Long ocean voyages deprived sailors of more than just fresh food. Quarters were damp and dirty, the air stuffy, and the drinking water stale. So when sailors reached land and their health improved, who was to say whether it was the good food, the fresh air or the pure water? The complexity of the problem also played to a deeply held belief of eighteenth-century medicine. This belief was opposed to today's view that many diseases have specific causes, be it a viral infection, a genetic mutation or a vitamin deficiency. At the time, doctors believed that any one disease can have multiple causes; you might get scurvy from unclean water, damp air or dirty quarters, and the actual cause would depend on your constitution. And if diseases have multiple causes, so the thinking went, they must also have multiple cures. For this reason, the medical establishment frowned on what was called a 'specific' – a remedy for a single disease that worked for all people. If a physician claimed to have found one, accusations of quackery were not far away.

This deeply embedded belief was also the reason why Lind's controlled trial did not cause a stir: such a trial was ideal to find specific cures for scurvy, but that's not what doctors were looking for. And in this respect, Lind was not far ahead of the pack. In fact, he was right in the middle. That's because – get ready for this – he did not believe in a specific treatment for scurvy either. He wrote in the *Treatise*, 'There is not in nature to be found an universal remedy for any one distemper.'[16] And because of this entrenched belief, Lind himself did not fully appreciate the significance of his own trial, and devoted little of the *Treatise* to it.

Eventually, common sense would win against medical prejudice, but it would take almost another half century. A tipping point in favour of citrus fruit came from a 1793 scurvy outbreak in the British fleet. The outbreak was quashed when lemon juice brought in from port helped to heal suffering sailors. A few incidences like this persuaded the Admiralty that citrus fruit were specifics, even though many physicians stuck to their old beliefs. By 1800 it was official policy to supply all ships in the fleet with lemon juice, and in 1867 the practice was also forced on civilian ships. It is the root of the American nickname 'limey' for the British.

Lime juice could cure scurvy, but for several more decades nobody knew why. The final answer would not arrive until 1927, when the Hungarian biochemist Albert Szent-Györgyi isolated a molecule he called *ascorbic* acid, after others had shown that it was the ultimate specific antiscorbutic. Today we know it as vitamin C.

Effective antiscorbutics were held back by prevailing scientific belief, a very human reason why crucial innovations remain undervalued. And this human element does not only plague the innovations of medicine. The next example – refrigeration – highlights another human factor that can render innovations not just dormant but comatose.[17]

Most refrigerators rely on a liquid refrigerant, its ability to absorb heat when it vaporises, and to release that heat when it condenses again. Today's household refrigerators use an electric compressor that helps condense vaporised refrigerant. This compression also helps the refrigerant to liberate heat that it previously absorbed from inside the refrigerator. Such refrigerators emit a constant humming whenever the compressor's electric motor is working.

A completely different, arguably better, refrigeration principle is embodied in the absorption refrigerator. This machine does not directly manipulate the refrigerant's pressure but its temperature. It heats a refrigerant to evaporate it, which later allows the refrigerant to release stored heat into the environment by condensation.

Absorption refrigerators occupy small market niches today, for example in recreational vehicles or off-the-grid living, where electricity is not available, and where a gas flame is used to heat the refrigerant. But in the early twentieth century, gas-powered absorption fridges were popular. They competed with electrically-powered compression models. That's no surprise, because lacking a compressor and its moving parts, they are easier to maintain. Also, they are silent, whereas compressors are noisy. And in the mid-1910s, when the first household refrigerators were being developed, more households were serviced by gas than by the electricity needed for compressor models. At the time, absorption refrigerators were even called 'the common sense machine'.[18] Nonetheless, compression models gained the upper hand. The reason had little to do with advantages like their higher energy-efficiency, but everything with powerful companies and their marketing muscle.

The first household refrigerator that reached the market in 1918 happened to be a compression model called the Kelvinator, created by an eponymous company and sponsored by the deep pockets of General Motors. Soon thereafter, dozens of companies jumped on the refrigerator bandwagon. Most notable among them was General Electric (GE), because it was a central player in the electrical industry, with a vested interest in all electrical technology, and for good reason: increased demand for electricity would help GE sell power-generating equipment. And so General Electric not only improved compression refrigerators, but supported them with huge advertising campaigns, including one with a neon sign visible from three miles away, a fridge shipped on a submarine headed to the North Pole and an hour-long infomercial with Hollywood stars.

In contrast, the companies who developed absorption refrigerators were fewer, smaller, had shallower pockets, and needed to ally with gas utilities, which were under financial strain from competing with the rising electrical industry. By 1940, General Electric and a few other companies wedded to the electrical industry – Westinghouse, Kelvinator and Frigidaire – had cornered the market. As historian Ruth Schwartz Cowan puts it, 'We have compression, rather than absorption refrigerators in the United States today not because one was technically better... but because General Electric, General Motors, Kelvinator, and Westinghouse were very large, very powerful, very aggressive, and very resourceful companies...'

Marketing was just as fateful for many other innovations. Among them is the vacuum cleaner. The late nineteenth century saw the development of centralised home vacuum cleaners, in which a single power unit drives vacuum hoses connected to vacuum outlets in each room. Nonetheless, today we mostly use portable rather than centralised vacuum cleaners, because successful companies like Hoover favoured them. Likewise, we use washing machines with a rotating drum, because Maytag aggressively promoted that design.

Be it marketing power or prevailing opinions that render some innovations dormant, the ultimate cause is social learning – or a lack thereof. The same force that spreads tool innovations through animal populations is also crucial to spreading human innovations. And when social learning is hindered or delayed for whatever reason, so is an innovation's success. But this does not mean that social learning is all that matters for this success. Some innovations remain dormant for similar reasons evolutionary innovations do, and these reasons have little to do with human foibles. Consider the remarkable case of a Swedish man named Arne Larsson, and the breakthrough innovation that helped him achieve a long life against impossible odds.

In 1958, Larsson was thirty-three years old and a desperate man. A viral infection had left scars on his heart that caused an abnormally slow heartbeat. As a result, Larsson suffered up to thirty dangerous fainting spells known as Stokes-Adams attacks every day. Each of these attacks could be deadly unless he was resuscitated in time.

The timing of Larsson's heart was clearly off, but he showed impeccable timing in other ways. In a Swedish hospital where he had already spent six months, the surgeon Åke Senning and the engineer Rune Elmqvist were experimenting with cardiac pacemakers. They wanted to be the first to develop pacemakers that could be implanted into patients, to feed arrhythmic hearts with the tiny electric pulses they need to keep beating. Larsson also had a formidable ally in his spunky wife who was almost as desperate as Larsson himself, because she was frequently the one who had to resuscitate him. She would pester Senning and Elmqvist for help, and when they told her that they simply did not have an implantable pacemaker yet, she told them simply to make one.

The history of cardiac pacemakers arguably began much earlier, in the late eighteenth century, when French anatomist Marie-François-Xavier Bichat jump-started the hearts of deceased people with electric current. That would have been a marvellous medical advance, an immediate life-saver for many patients, except for one problem: Bichat's patients lacked heads. They were decapitated victims of the ongoing French revolution.[19]

From proofs-of-concept like these, it would still take more than another century until the first efforts at cardiac pacemaking succeeded. A milestone year in this delayed success story was 1932, and it is linked to the American cardiologist Albert Hyman. Hyman was less interested in managing chronic heart disease than in treating acute, once-in-a-lifetime problems. For example, he wanted to jump-start the quiescent hearts of stillborn babies. Hyman is credited with the name artificial pacemaker that is still used today, but the contraption he invented had little in common with today's pacemakers. It was the size of a sewing machine, with a hand-cranked motor that generated DC power to be transmitted via a needle into a patient's heart.

You would think that the medical establishment would welcome Hyman and his revolutionary invention with open arms. But no, the medical community dismissed Hyman's pacemaker as a mere gadget, and he was even accused of interfering with God's work. In other words, Hyman's pacemaker did not catch on, and it was certainly not for lack of trying. More than a decade after his original invention and well into World War II, Hyman tried to persuade the US Navy to use his device, but without success. In the end, the field of pacemaker research lay fallow for another twenty years, until it began to yield an abundant crop in the late 1950s.

That decade saw two breakthroughs. The first of them occurred just one year before Arne Larsson's long-term hospitalisation in Sweden, when C. Walton Lillehei, a cardiac surgeon in Minneapolis, teamed up with Earl E. Bakken, an engineer, one-time TV repairman and inventor working out of his garage. Lillehei had pioneered open-heart surgery to correct congenital heart defects in children. Because some of his patients would develop abnormal heart rhythms until they had recovered completely, he was keen to support their hearts temporarily with an artificial pace maker. Enter Bakken, who realised that metronomes and pacemakers solve the same problem – how to keep a beat. So he co-opted a design for a small metronome from an electronics magazine and built a pacemaker from it. Powered with a nine-volt battery, his device was small enough to be strapped to a patient, with a wire running through the chest wall to the heart.

But Lillehei and Bakken had not solved one crucial problem: a permanent wire-sized hole in the skin is a giant gateway for bacteria and the life-threatening infections they cause. The solution – obvious, but perhaps only in hindsight – is to implant the whole shebang, pacemaker and wires, into a patient's body. But that requires a very small device and, just as important, non-toxic materials stable enough to last for years inside the body. To solve this problem, the Swedish team used silicon transistors, which were not just brand new then but also tiny, and could be powered by a small rechargeable battery. It also helped that the chemical company Ciba-Geigy had just developed a novel kind of biocompatible epoxy-resin called Araldite, which they used to seal the whole package. The end result was a cylindrical device about the size of a hockey puck, which they transplanted into Larsson's abdomen.

Sadly, this first pacemaker lasted only eight hours. However, Senning replaced it with a second one that lasted a full week.

Fig. 9 An early example of a pacemaker

Larsson would receive replacement after replacement, and survived for many years. When he died in 2002 at the age of eighty-six, not from heart disease but from melanoma, he had undergone twenty-five procedures to replace failed pacemakers with smaller, better and safer ones. He had also outlived both Senning and Elmqvist.[20]

Curiously, the original inventors did not see the success of their pacemaker coming. Elmqvist said, 'I must admit that I had regarded the pacemaker more or less as a technical curiosity.'[21]

But by the time Arne Larsson died, some three million people were walking around with pacemakers in their chests, and today US doctors implant more than 100,000 pacemakers every year.[22]

The dormancy of pacemaker technology between Hyman's hulking monster and the mid-1950s had less to do with obstacles to social learning than with the speed of technological development. A fully implantable pacemaker needed miniaturisation that could not be achieved with bulky vacuum tubes, but only with small and power-efficient transistors, and these would not be invented until 1948, sixteen years after Hyman's invention. What is more, implantation also required batteries that were both small and long-living, and biocompatible materials like Araldite. In other words, the implantable pacemaker had to wait because not just one but several necessary technologies had to be invented first.

Merging streams of innovation were also essential for the success of another technology: hydraulic fracturing or 'fracking', which helps to extract oil and natural gas from rock. Loathed by environmentalists because it prolongs our dependency on fossil fuels, produces chemical waste, contaminates drinking water and even creates minor earthquakes, fracking has helped revolutionise the fossil fuel industry by opening massive reservoirs of natural gas and oil.[23] Its roots go back to the mid-nineteenth century when drillers were 'shooting' unproductive oil wells by igniting gunpowder deep inside a well. The resulting explosion would not only warm the heart of every boy and man, creating a deeply satisfying geyser of oil and water that shot out from a borehole, but its pressure would also fracture the rock near the well, creating fissures in the rock that allowed oil or gas to seep into the well to be extracted.

But from that embryonic state of fire and fury, it took until 1947, almost another century, for fracking itself to be born.

That's when Floyd Farris from Stanolind Oil realised that fluid pumped into a well at high pressure could achieve the same result as an explosion.[24] Fracking would soon be used routinely, but its breakthrough success came only several decades later, when another technological process known as horizontal drilling had matured. Horizontal drilling is remarkable, because it allows a drill bit to change direction while grinding away at rock more than a kilometre underground. The drill bit describes a ninety-degree arc from vertical to horizontal, creating a well that runs parallel to the surface. Horizontal drilling is important, because most reservoirs of gas and oil lie within sheets of rock that are much wider than they are thick. The longer a well runs inside such a sheet, the more oil and gas can be extracted.

In the 1980s the innovation streams of fracking and horizontal drilling merged. Their merging enabled what is called 'massive hydraulic fracturing', where more than ten million litres of water mixed with more than a hundred thousand litres of chemicals are rushed down a well. The resulting pressure cracks open rock deep inside the earth, and releases copious amounts of gas and oil.

I am describing examples from scurvy to fracking in some detail to make a simple point: the reasons why an innovation has to wait for success are enormously diverse, so diverse that it's not easy to see what they might have in common. But that's as far as examples can take us. Unfortunately, they cannot reveal broader patterns in the history of human innovations. After all, they are just singular, unique episodes in this history. Most importantly, they cannot tell us how frequent dormant innovations or discoveries are to begin with. Are they perhaps just rare exceptions to some general rule?

Fortunately, a priceless source of data can help answer this question, at least for the history of science. This source is the paper trail of scientific publications, a centuries-long, indelible, time-stamped record of discovery. One can use this record to study the universal currency of recognition and success in science – the scientific citation.

When one scientist cites another, she pays an intellectual debt to that earlier work. The more citations, the greater the debt. In other words, citations can quantify a work's influence. They can help predict which scientists will reap coveted prizes like the Nobel.[25] They help universities identify the most promising job candidates. And they are the subject of a specialised branch of science called scientometrics, also known as the science of science.

Using scientometrics to determine a scientist's value can be problematic, especially for outstanding young scientists whose work may need time to be recognised. However, when applied not to individuals but to entire fields of research, scientometrics can help enhance the acuity of hindsight and identify broad historical patterns of influential work. It can be especially useful to screen electronic databases with millions and millions of papers for dormant discoveries. These are embodied in papers accumulating few or no citations for years or decades, followed by a burst of citations when they become popular. Scientometrics uses 'sleeping beauty' as the technical term for a paper whose recognition is delayed. It studies how the 'heartbeat' of such papers – the number of citations per year – quickens when they are 'awakened'.[26]

One 2015 study of a whopping twenty-two million scientific papers did just that. It defined and quantified a sleeping beauty coefficient B that encapsulates in a single number how long a discovery remains unrecognised, how few citations it receives during this time, how rapidly its citations spike once it becomes awakened, and how many citations it receives during that delayed citation burst. At one extreme are publications with the highest values of B. They describe discoveries that have been neglected for many decades, only to experience a huge surge of popularity when their time has come.[27] Conversely, publications where this coefficient B equals zero are no sleeping beauties. Their citations spike right after publication – immediate recognition – only to ebb afterwards, a typical pattern for many publications.

The authors of this study wondered whether sleeping beauties stand apart from all other publications, forming a cluster, perhaps small, of publications that are easily distinguished and counted. But that's not the case. Beauty coefficients vary over a continuous range, from many publications with tiny B to fewer with intermediate B, and even fewer with large B. Also remarkable is that the distribution of B is nothing like the familiar bell curve or normal distribution. Instead, it is what's known as a fat-tailed distribution, because its 'tail' – publications with a high B that are ignored for a long time – encompasses many more publications than the tail of a bell curve. In other words, there are many more long-dormant and suddenly roused discoveries than expected by a bell curve. Sleeping beauties are not rare exceptions in science.[28]

Some of them are well-known, such as the work of the monk Gregor Mendel. He toiled over genetic crosses with pea plants in the Austrian monastery of Brno, at about the same time that Charles Darwin laboured over the *Origin of Species by Means of Natural Selection*. Darwin's theory had a gaping hole, because Darwin did not know how inheritance worked, nor why children resemble their parents.[29] Unbeknownst to Darwin, Mendel laboured to fill this hole, crossing thousands of pea plants whose seeds and flowers varied in shape, colour and texture. When Mendel meticulously recorded these traits in their offspring, he found something curious: some of these features pass from parent to offspring completely intact, over multiple generations, as if indivisible atoms of information were handed from generation to generation. Mendel described this discovery in an 1865 paper that would have been accessible to Darwin. It was mentioned in books that Darwin had read, and Mendel himself may have sent Darwin a copy. However, we have no evidence that Darwin was aware of it. And Darwin was not alone in overlooking Mendel's insights. Everybody else did too.[30] For more than three decades, that is, until the Dutch botanist Hugo de Vries rediscovered Mendel's paper in 1900, and triggered the genetics revolution which culminated in applications like genetic engineering.[31]

For every well-publicised rediscovery of a Mendel, there are many others that fly under the public's radar even after their awakening. Among them is a 1953 study which reported that some cells synthesise fat-like molecules with the forbidding name of phosphoinositides when exposed to neurotransmitters – molecules that help brain cells communicate. The world didn't seem to know what to do with this discovery, because the study remained obscure for twenty years before being roused. By that time, it had become clear that the release of these molecules is part of a complex molecular relay chain known as calcium signalling.[32] And calcium signalling is central whenever cells need to communicate. It helps cells divide, muscles contract, embryos develop and neurons fire.

Just as obscure is a 1971 article in which a Japanese team described a new disease of the retina, only to be ignored for more than a decade. Its awakening came when others not only rediscovered the disease, but found that it was caused by a herpes virus infection.[33]

These and many other arcane discoveries were premature for a common reason that we have already encountered: their success had to await other developments. The jargon of scientometrics calls these developments 'princes'. They awaken a dormant publication, which then experiences a burst of new citations.[34] For phosphoinositides the prince was a discovery about cell communication. For retinal disease, it was the discovery of a cause, which is important for diagnosis and treatment. In other words, the success of scientific discoveries, like that of technological innovations, can require multiple building blocks. Think of them like parts of a model airplane that you assemble from a kit – the plane can fly only when all its parts are in place.

These building blocks often come from other disciplines. In fact, three quarters of sleeping beauties are awakened by developments from another discipline,[35] a reminder that major scientific advances often come from cross-fertilisation between different fields.[36]

Some of these building blocks need not even be scientific discoveries. They may come from outside science. When Alexander Fleming discovered penicillin in 1928, it would remain a laboratory curiosity for more than ten years. The reason: discovering an antibiotic-producing mould is one thing, and turning that discovery into a useful drug is another. Nobody showed much interest at the time, so Fleming himself gave up penicillin research in 1929. When interest in penicillin eventually arose, one of the reasons was entirely non-scientific: World War II was raging and thousands of soldiers dying from infected wounds.[37] But even with that motivation, the isolation, clinical testing and mass-production of penicillin required a multi-year effort and interdisciplinary teams from the UK and US. Only in March 1942 would the first US patient be treated with penicillin.

A pattern like this also exists for discoveries that appear initially useless, but given enough time develop surprising applications. They seem especially abundant in mathematics, perhaps because mathematicians, like many artists, proudly pursue work with no immediate utility. Here is what eminent

Fig. 10 A poster hailing the advent of penicillin

British number theorist G. H. Hardy wrote in 1940, towards the end of a successful career: 'I have never done anything useful. No discovery of mine has made, or is likely to make, directly or indirectly, for good or ill, the least difference to the amenity of the world... Judged by all practical standards, the value of my mathematical life is nil.'[38] His fellow number theorist Leonard Dickson reportedly expressed this sentiment more succinctly when he said: 'Thank God that number theory is unsullied by any application.'[39] Indeed, for centuries number theory had been considered one of the purest field of mathematics, because its applications were so few. But neither Hardy nor Dickson knew what was about to hit their cherished part of the ivory tower. With the rise of computers, applications of number theory exploded, and by the late twentieth century they had helped solve problems in physics, finance, biotechnology, but most of all computer science. Today they are essential for the cryptographic algorithms that ensure secure communications throughout the internet.[40]

Another branch of mathematics called group theory also started out as pure mathematics, but became important in particle physics and crystal chemistry starting in the 1920s. And likewise for the mathematical foundations of physics, which include James Clerk Maxwell's 1865 complex equations linking light, electricity and magnetism. Maxwell was not interested in any applications, and neither was Heinrich Hertz, who first proved thirty years later that Maxwell's equations revealed more than just mathematical truth, when he created and detected radio waves in 1887. Not until the 1890s, thirty years after Maxwell, would the first killer app emerge, when Guglielmo Marconi began to develop wireless telegraphy and radio.[41]

Even more remarkable than these examples are multiple publications that appeared as early as 1893 and that described statistical methods to analyse data. They are remarkable because they were barely noted in the twentieth century – some of the earliest ones remained dormant for the *entire* twentieth century – until they skyrocketed to prominence in the early twenty-first. The reason: 'big data' and powerful computers that can let these

methods shine.[42] Likewise for deep learning, a highly successful branch of artificial intelligence that uses neural networks to
recognise speech, classify images and identify people from their
faces. The ideas behind it go back to the 1960s and were well
developed by the 1980s, but needed massive data sets that did
not emerge until the twenty-first century to reveal their power.[43]

Just like the innovations of biological evolution, many human inventions and discoveries lie dormant, except that better historical records help us to pin down the reasons for their
dormancy. And these reasons are enormously diverse. Some of
them may involve little more than dumb misfortune, like Mendel living in an academic backwater. Others include entrenched
prejudice, inertia to change and marketing power. Still others
are scientific discoveries or technological developments that lie
in the future. Because every one of them is unique, they seem to
have little in common. But deep down, they do share something,
and that something is crucial: all of them lie outside the creator's
control. And this commonality, which runs like a red thread all
the way back to the origin of life, may also hold a few lessons for
human creators today.

10

Sleeping beauties

CAN HUMAN CREATORS – ESPECIALLY struggling ones – find any comfort in the themes of this book? Before suggesting an answer, I'll allow the main themes a brief lap of honour to examine what they teach us about our own success and failure.

Here is nineteenth-century British historian and politician Thomas Macaulay on the first theme – multiple discovery: 'The sun illuminates the hills, while it is still below the horizon; and truth is discovered by the highest minds a little before it becomes manifest to the multitude. This is the extent of their superiority.'

These lines reflect a dim view of genius and its role in science and technology. A genius is merely a few steps ahead of the pack. That view is supported by the sheer abundance of multiple discoveries. And it's actually even worse than Macaulay writes: the work of that genius often remains unrecognised until the rest of the world has caught up.

I mentioned some examples but skipped many others. For instance, several early and unique discoveries in chemistry and physics turned out not to be so unique when the unpublished research records of Henry Cavendish came to light after his death in 1810. Among them is Ohm's law relating electrical current and voltage, and the finding that combusting hydrogen produces water. Cavendish had been there first. Likewise, the discoveries of multiple mathematicians appeared to be singletons, until they were found in the private notebooks of eighteenth-century mathematician Carl Friedrich Gauss.[1]

Multiple discoveries are the rule, singular discoveries the exception, and so is the role of the singular genius. If received wisdom tells us otherwise, it's because history is written by the winners (and by those who excel at self-promotion).

Singular genius seems to have just as little place in other species, where innovations like termite-fishing in chimpanzees, sponge-fishing in dolphins and potato-washing in macaques were all discovered multiple times.

And the same holds for the mindless process of biological evolution. The best known examples involve complex organs like the eye, which evolution discovered multiple times, and in organisms as different from us as molluscs, flies and jellyfish.[2] But many less well-known examples are no less dramatic. One of them involves the snake *Enhydrina schistosa*. It is also known as the hook-nosed sea snake, because the scales on its head are arranged into a beak-like shape that may help it feed on spiny fish. It is famously aggressive and produces a lethal toxin that accounts for most deaths from sea snake bites in Asian and Australian waters. Uniquely deadly, it would seem. However, researchers who reconstructed its evolutionary past from its genes found that it is not one species but two, an Asian and an Australian one. The two species independently evolved an uncanny resemblance, from the beak down to their aggressive behaviour, and to the chemical composition of their nasty venom.[3]

Such convergent evolution reaches all the way down to the submicroscopic level, to molecules like caffeine that help plants to fend off their enemies. Evolution has discovered caffeine – the exact same molecule, with the exact same twenty-four atoms, and their exact same arrangement – at least three times, in coffee, tea and cocoa plants.[4]

Fig. 11 A hook-nosed sea snake

Merton's multiples are so frequent in nature that Cambridge palaeontologist Simon Conway Morris argues that none of life's innovations are unique.[5] 'I'll always say "show me anything which has only evolved once" and I'll try and jump up and say "no I can give you another example"' is how he put it in a 2015 interview with the *Independent*. Even extra-terrestrial life, he argues, would look much like us.[6] Not so fast, say other scientists, who point out that we need not look to other planets to find singletons of evolution – our own harbours bizarre species like the chameleon, the desert plant *Welwitschia* and the duck-billed platypus.[7]

But here again, the appearance of uniqueness can deceive. Thanks to the mind-boggling time scales of evolution, our knowledge about the history of life is riddled with holes. Who can be sure that a new way of life, body shape or molecule has not evolved more than once, perhaps countless million years ago, but that only one has survived to this day, the eons having erased all traces from those unsuccessful trials that went extinct? Just consider the earliest mammals from chapter 2 that resembled tree-shrews, beavers and flying squirrels. Their life-styles had been discovered by evolution, went extinct again and had to be rediscovered – some more than once – before they eventually succeeded, many million years later.

The debate about whether *any* natural or technological innovations are truly singletons will continue. But no matter how it ends, the mere abundance of multiples teaches us that innovation usually comes easily to both nature and culture. We don't know all the reasons, but an important one comes from a characteristic they share, namely that many innovations are not built from scratch. Instead, they co-opt the old to create the new.

When Johannes Gutenberg revolutionised communication technologies in the fifteenth century by inventing the printing press, he combined the screw press and moveable type. Both had been around for centuries. Screw presses had been used for wine-making since Roman times. Moveable type, although of Chinese origin, had been mentioned in European records long before Gutenberg's time. Likewise, the wheelbarrow, whose European origin dates to the twelfth-century, combines the lever and the wheel, two technologies that originated in antiquity.[8]

A special kind of such co-option does not use existing technology but existing knowledge. The innovation of radar, for example, is based on the knowledge that metallic objects can reflect radio waves, just like mirrors can reflect light. Nature has inspired many such analogical innovations. Take the design of the Eastgate Centre, a shopping and office complex in Harare, Zimbabwe. Its cooling system was inspired by termite mounds, which are cooled passively as warm air rises through them and draws cooler air behind it. Likewise, the discovery of Velcro was inspired by the burrs that clung to Swiss inventor George de Mestral and his dog during their forest walks.

Nature, not culture, is in fact the true master of such co-opting and combining. When bacteria modify their metabolisms to harvest energy and mine materials from new molecules – including man-made toxins like antibiotics – they swap snippets of their genomes and try ever-new combinations of genes and enzymes, until they find one that can digest a new food source. And DNA itself is the ultimate combinatorial innovation engine. Its four letters can be re-combined and mutated into ever new genes and genomes, a process that brought forth all the millions of species alive today.

Frequent co-option also hints at the tremendous potential of the old to serve new purposes. And it leads to another main theme: many innovations are born in a latent, quiescent state. This theme manifests itself most clearly whenever changing the old is not even necessary to meet a new challenge. That is the case in metabolism and its molecular road network. Metabolism can digest molecules that it has never before encountered, as long as they can travel one of its many chemical pathways, a bit like a street that can be used by people for whom it has not been built, as long as it passes close to their home.

Innovations like these also slumber in promiscuous enzymes that can do a dozen jobs, even though evolution shaped them for only one, such as building a single chemical weapon. They also exist in the very circuits of our brain that enable science and technology. The circuit that makes reading possible did not evolve to recognise letters and words, but objects and landscapes. Letters and words, however, have much in common with other objects, and so this circuit, older than humanity itself, was ready to take on reading when reading was needed.

To blossom, innovations like these may not require any change in an old circuit, molecule or technology, but they do require a change in the environment. Such innovations embody, in its purest form, the last and most important theme of this book: an innovation's success depends on the world around it.

That's why innovating successfully can require patience. All those sleeping beauties we encountered needed to wait anywhere between years and millions of years until the world was ready for them. And their waiting can take many forms. Some innovations survive in the small niches they fill at the margins of culture or nature. They include grasses, ants and birds, which eked out a living for many million years until their times had come and they experienced explosive success. They also include many human innovations, and even more that are still waiting, like the gas-powered absorption fridge, which may become wildly successful one day, continue to scrape by, or eventually become extinct.

Other innovations persist because their new job is a by-product of an old job. Among them are the crystallins, some of which accelerate chemical reactions but can also help diffract light in the eye, because they don't clump at high concentrations.

Yet other creative products persist unused. In culture, they include forgotten novels, ignored symphonies and overlooked patents. In nature, they include the hundreds of recently birthed genes that large genomes harbour.[9] Each such gene is a solution in search of a problem. If the problem never arises, the gene solving it will eventually disappear. Mutations will slowly turn its DNA to gibberish or delete it from the genome, just like a forgotten manuscript will eventually perish, be it with a bang in a fire, or with a whimper in the recycling bin.

And then we have those innovations that actually go extinct, like early mammalian lifestyles or early cures for scurvy. They have to be rediscovered again later. If innovation were any harder than it is, they would have been lost forever.

Even more diverse are the reasons why sleeping beauties sleep. When mathematical creations like the linear algebra of

Hermann Grassmann are ignored for decades, we can blame the ever-elusive zeitgeist. This zeitgeist can also stymie discoveries that are about utility rather than beauty or truth. Among them is the first discovery of radar by the German Christian Hülsmeyer, who envisioned radar as a technology that could save lives in both war and peace. But others did not share his vision; radar was rejected by the German Navy as unnecessary, and had to be rediscovered twenty years later by the Germans and the British.[10]

Politics is part of this zeitgeist. The Korean script Hangul was invented in the fifteenth century, and expanded the world of writing from a small elite literate in Chinese to the general population. But this expansion had an unintended consequence: by allowing the broad masses to express themselves, this new technology threatened the ruling class. Sure enough, when Hangul writing was used to criticise the king, the innovation was banned for half a century.[11]

Human foibles are not all that matters for success or failure, as some of the technological innovations we encountered illustrate. Fracking was able to revolutionise the oil industry only once horizontal drilling technology had matured. The cardiac pacemaker could not become practical until transistor-driven electronics, small batteries and non-toxic materials allowed it to become implanted into a patient. Penicillin could revolutionise medicine only through technologies to mass-produce it, which took more than a decade to develop.

The innovations of biological evolution make a similar point. They tell us that being born in the right time and place matters even in a world free of humans. Ants had to wait for tens of millions of years, until flowering plants had spread and formed large and diverse forests, before they were able to diversify into more than ten thousand species. Mammals had to wait until most dinosaurs had gone extinct – all but the early birds in fact, which also had to wait until their dinosaur brethren were gone. And the grasses and cacti had to wait for the climate to change, because they could succeed only on a dry planet.

All these reasons for delayed success – climate change, incumbent technologies, asteroid impacts, power politics, atmospheric oxygen, personal prejudice, plant diversity, political upheaval – have so little in common that they bring to mind the old saying 'history is one damned thing after another'. But they do have one thing in common: they do not originate in the innovation itself, but in the world outside.

And they lead to a troubling conclusion: no innovation succeeds on its own merit. This may be hard to swallow for those of us who believe that creative success arises from intrinsic genius. We may think that a novel is so moving that it should be an instant bestseller, that a scientific discovery is so obviously groundbreaking that it should be widely acclaimed, and that a new medicine is so useful that humanity could not possibly do without it. But for no such innovation does any inner characteristic or 'quality' suffice for success. The novel may speak to a narrow audience, the scientific discovery may need to overcome prevailing opinions, and the new medicine may await a method for mass manufacture.

In addition, the very notion of inner quality is often elusive and ambiguous, a point well made by some of nature's innovations. Among them is an ancient molecular innovation that helps save the lives of many people living in malaria-infested regions of Africa. It causes a single amino acid change in the haemoglobin protein that helps our blood transport oxygen. And this single mutation – a change in a few minute atoms within a huge protein – has a tremendous consequence: it prevents the tiny protozoan that causes malaria from spreading inside our body. People who carried this mutation in malaria-exposed regions remained healthier than others. Over many generations, thanks to natural selection, their descendants increased in numbers.

Today, more than twenty percent of people in some African populations carry this mutation.[12]

A clear-cut case: a mutation that defends our body against a disease, a good thing all-round. But to most students of biology – even many secondary school students – this mutation is not known as an innovation. It is known as the cause of a disease just as devastating as malaria: sickle cell anaemia, named after the abnormal sickle-shape it causes in red blood cells. These cells cannot transport oxygen, thanks to the very same mutation that protects against malaria. Just as we carry two copies of each gene, we also carry two copies of the genes encoding haemoglobin. If only one of these copies is mutated, we have only mild anaemia but are protected against malaria. If both are mutated, we suffer life-threatening anaemia.[13]

The sickle cell mutation is not the only such double-edged genetic innovation. Another one revolves around mutations that protect against tuberculosis, a disease that has been widespread for at least fifteen thousand years, and caused at least twenty percent of all European deaths between the sixteenth and twentieth centuries. Unfortunately, these mutations too are Janus-faced. Some of them also cause cystic fibrosis, among the most common genetic diseases in Europeans, and just as deadly as sickle cell anaemia.[14]

More generally, many DNA mutations can be helpful or harmful, depending on when and where they occur. Geneticists call this frequent phenomenon a genotype-by-environment interaction.[15] It blurs the distinction between good and bad, at least in the realm of biology. If beauty is in the eye of the beholder, so is utility.

* * *

The Dutch painter Johannes Vermeer has the dubious honour of being among the longest neglected creators of the last millennium. He was born in 1632 and his output as a painter – only some fifty paintings in his life-time – was not great. It was especially modest considering that he had eleven children to feed. Vermeer was respected by his colleagues, but that respect did not translate into sales, and when he died in 1675, during a time of war and economic crisis, he left his wife and children mired in debt. His work quickly fell into oblivion and remained there for almost two centuries. That's how long art historians and critics ignored him.

The tipping point came in 1866, when art critic Théophile Thoré-Bürger published a catalogue of Vermeer's work that went viral, as we would say today. As if to make up for lost time, Vermeer's style became not only wildly popular and widely imitated, but he was quickly catapulted to Olympian rank. Today his work even permeates popular culture, through novels and movies such as the best-selling *Girl with a Pearl Earring*.

Vermeer belongs to a huge club of neglected creators that eventually broke through. Another member is American naturalist and part-time hermit Henry David Thoreau, whose 1854 book *Walden* described his two-year seclusion in the woods around Walden pond.[16] Only modestly successful on publication, and going out of print six years later, the book would experience a long dormancy from which it was awakened by twentieth-century environmentalism, to become one of America's most celebrated literary works.

Another case in point is John Keats, whom I have already mentioned. His was not a bad time to be a poet, because popular narrative poems like Lord Byron's *The Corsair* could sell ten thousand copies in a single day, and a volume of poetry by a fashionable poet could fetch more than £3,000 – some 250,000 US dollars in today's currency. But that kind of success would elude Keats. His first volume of poetry sold so poorly that a close friend remarked that it might as well have been published in

Timbuktu. One reviewer described his narrative poem *Endymion* as 'drivelling idiocy'. It would take decades after his death until his work became recognised.

The club arguably even includes Johann Sebastian Bach. He was well respected in his time as a keyboard performer and composer, but surprisingly, given his godly status today, not beloved for his choral works. His contemporaries had simpler tastes in song. As late as 1800, half a century after Bach's death in 1750, one critic described the complexity of Bach's chorals as 'absurd'. What is more, Bach's music was considered old-fashioned and went out of style after his death. For many decades the senior Bach's compositions were used as mere tools to teach harmony and counterpoint. They might eventually have been forgotten altogether, were it not for an 1802 Bach biography that helped refamiliarise the public with his work. Another tipping point was a celebrated 1829 performance of the *St. Matthew Passion*, which launched a wave of popularity that engulfed the general public and continues to this day.[17]

Creators like these and many others – El Greco, Emily Dickinson, Herman Melville, Vincent van Gogh – did not benefit from radio and television broadcasts that spread information to millions of people. Neither did they benefit from the internet that spreads information to billions. So perhaps that explains their delayed recognition. Perhaps we all would recognise the next Keats instantly if we all just knew about him. Perhaps. But many forgotten creators were not unknown, so that can't be the whole explanation. They often had a small circle of devoted friends who promoted their work before it became forgotten.

And even in today's world, where information instantly spreads across the globe, and where the internet offers electronic immortality, oblivion does not seem to have gone out of style. Instead it seems to be booming. For this we can thank the crushing tsunami of immersive information sweeping over us every day, and overwhelming our senses.

One sign of this boom is the burgeoning literary category of 'lost classics' or 'neglected books'. New York Review Books Classics is a publisher dedicated to reissuing out-of-print books whose greatness is nearly always vouched for by literary luminaries. It has reissued hundreds of previously forgotten titles. Other titles that never made a dent are published in imprints with names like Second Chance Press and Lost American Fiction. And many more appear in web-based collections and anthologies like Unknown Masterpieces or They Died in Vain.[18] By these accounts, forgotten yet memorable writers must number in the thousands. Stephen Marche, a contemporary writer, lends a voice to them when he says: 'All of our little streams pour out into the ocean of total uncaring… persistence may be the one truly writerly virtue, a salvation indistinguishable from stupidity. To keep going, despite everything. To keep bellying up to the cosmic irrelevance. To keep failing.'[19]

The question of why artistic quality, and yes, genius, is not always instantly and universally recognised has been on the mind of countless struggling creators. Some, like philosopher Hannah Arendt, asked whether 'unappreciated genius' wasn't just the 'daydream of those who are not geniuses'. But this cannot be the whole truth, as the lives of Vermeer, Thoreau, Keats and many others prove.

The key to the truth is that sleeping beauties do not just exist in the arts, but everywhere in the human realm – technology, science, maths – and far beyond it, all the way back to life's origin. Their ubiquity shows that innovations cannot stand on their own. They must be born into a world that's right for them.

This basic but universal fact immediately prompts two prescriptions to help human creators find success. The first is to listen to the world and find out what it wants. That's exactly what thousands of authors, screenwriters, publishers, movie producers, inventors, composers, directors and many other professionals do when they strive to land the next blockbuster. In *Hitmakers*, a well-researched book about successes and failures, journalist Derek Thompson examines what distinguishes the hits from the flops. He identifies principles like MAYA – most advanced yet acceptable – the idea that successful products from bestselling books to musical earworms need to balance the new with the familiar.[20] These principles may be useful in hindsight, but about foresight, he concludes more depressingly, 'when it comes to predicting the future, ignorance is a club and everybody is a member.' Or more succinctly, in the words of a screenwriter, 'nobody knows anything'. Pop groups like Led Zeppelin, books like the *Harry Potter* series, TV shows like *American Idol*, and iconic products like the iPhone prove him right. All of them were predicted to be dead-on-arrival.[21]

If finding out what the world wants does not work, what about another tack? What about changing the world, creating the environment in which an innovation can succeed?

That's ambitious. The mere possibility is a human privilege and off-limits for other species – even if the tiny mammals had known that the giant dinosaurs stood in their way, they could have done little about it. And as it turns out, it is also usually too ambitious for any one human. Two examples will show what I mean.

In the time of Thomas Edison, fierce competition was under way to bring clean and safe lighting to households, companies and public spaces. Gas lighting was popular, but it was sooty, polluting and subject to deadly explosions. Edison and others who developed incandescent light bulbs knew that electric lighting was cleaner. But it could succeed only in the right environment, one where electricity could be mass-produced and distributed to thousands of customers. So they decided to create that environment.

During this time, thousands of companies begged Edison to install electric power plants just for their own use, but Edison had a bigger goal. He wanted to power entire cities.[22] So he set out to build an enormous coal-fired central power plant on Pearl Street in Manhattan, the first ever such project in the US. Aside from the enormous cost of real estate, this project also faced numerous technical challenges. They included building huge dynamos that generate electricity, and laying kilometres of electrical wires in a densely populated city. Gas utilities gleefully pointed to the dangers of electricity whenever somebody got electrocuted, so Edison decided to lay all those wires underground, which created its own problems. It took several years, but in 1882 the plant went live. Initially it served only four hundred Manhattan light bulbs, but this number increased to ten thousand within two years.[23]

The race to electrify the United States was on, but it would be a marathon rather than a sprint. While it took Edison only a few years to develop a public power plant, it would take half a century – and thousands of engineers, businessmen and politicians – to create the environment in which electricity could succeed.[24] As late as 1939, future president Lyndon B. Johnson, then a congressman, still struggled with powerful utilities to get rural Texas electrified.[25]

A similar struggle took place in transportation. We heard earlier about the sputtering start of wheels in Egypt and Mesoamerica, where ancient peoples like the Olmecs did not use wheels for transport. For success, wheeled transport needs the right environment, one that includes smooth surfaces like rails or roads. When US railroad companies emerged in the early nineteenth century, they had to create this environment from scratch, metre by metre of laid railroad track, first in the east and then across the entire continent, which took about half a century.

The car needs its own kind of environment, and creating that environment took no less time. As late as the early twentieth-century, when Henry Ford would sell millions of his affordable Model T automobile, one observer wrote that away from cities 'even the roads were off-road'.[26] It helped that Congress subsidised highways over railroads, because railroad companies were loathed for their monopoly on transportation. They had a chokehold on their customers, among them many farmers who needed to get their products to market. Additional ammunition came from the military, which argued that good roads were vital to help move troops and equipment in case of war. Add to that some shady machinations of the car industry, which would buy up local trolley lines to replace them with buses, and the car could more than just compete with the train. It eventually crushed the railroad industry.

Nonetheless, more than a century elapsed between the opening of the Baltimore and Ohio Railroad in 1830 and the building of the interstate highway system, which began in 1956 and was completed only in 1992. Once again, thousands if not millions of people were involved in this process. No one creator, not even a powerful tycoon like Henry Ford, would be able to bend history on this scale.

Neither changing the world nor predicting its desires are realistic prospects for any one person. So how else can a struggling creator benefit from the knowledge that sleeping beauties are universal, and that the success of her creations is beyond her power? The question brings to mind the old joke where a man tells a friend about his years in psychoanalytic therapy. The friend asks 'Did it help you?' 'No,' answers the man, 'but now I at least know why I'm screwed up.' Knowing the deep history of innovations may not help us change our place in the world, but it can help us make the best of it. It can help us draw satisfaction from creative work, even though we may be 'bellying up to cosmic irrelevance' forever.

For one thing, we have to accept that every creative product is just a ticket to life's grand lottery. Biologists will not be surprised, because applied to evolution's products, this view has been Darwinian orthodoxy for a century. When biologists call DNA mutations 'random', they mean just that – mutations are no more likely to create something new and useful than they are to create something new and harmful.[27]

And if it is hard to accept that the same principle applies to human creations, it helps to consult the irrefutable record of science and its millions of publications. This record sends a clear message by revealing a general pattern of success. This pattern not only exists all the way back to the nineteenth century, it also holds in fields as different as history, geology and mathematics. It reveals that the most successful scientists tend to be those who publish the most over their lifetimes. Nobel laureates, for example, write double the number of publications than other scientists. What is more, scientists in general produce their most important work in those years where they publish the most. Scientometrics also dispels the urban myths that mathematicians and physicists are best when they are young and biologists when they are older. Instead, it proves that the chances of success do not change systematically over a lifetime, except with overall productivity. Psychologist and creativity researcher Dean Simonton calls this pattern 'constant probability of failure'.[28]

Just as in a lottery.

And this lottery – any lottery – has a simple prescription for its players. Buy as many tickets as you can afford – create as much as you can – and you will maximise your chances of success.

But just like in a lottery, most of your creations will fail.

That is a problem, but only if you are fixated on success. Curiously, many distinguished creators are not. This is what psychologist and creativity researcher Mihaly Csikszentmihalyi learned when he interviewed ninety-one eminent creators who included artists, writers, scientists, actors, musicians, architects and engineers.

Novelist Naguib Mahfouz says, 'I love my work more than I love what it produces', and writer Richard Stern, 'At your best, you're not thinking, how am I making my way ahead in the world by doing this?' When psychologist Donald Campbell talks to young people he recommends: 'Don't go into science if you will not enjoy it even if you do not become famous.'

Eminent creators also often have their own views about success. Neuropsychologist Brenda Milner declared that 'every new little discovery, even a tiny one, is exciting', and inventor Frank Offner said, 'I love to solve problems. If it is why our dishwasher does not work...or how the nerve works, or anything...it is so very interesting.' Nobel Prize-winning physicist Subrahmanyan Chandrasekhar stated, 'I have never worked in anything which is glamorous in any way...very often outside success is irrelevant, wrong, and misplaced.'[29]

What creators like these enjoy is the process itself rather than the finished product. In psychological jargon, what they do is autotelic: their work is an end in itself. They also possess in spades what psychologists call intrinsic motivation – creating for the joy of it, because it is satisfying in itself. Psychologists also know that intrinsic motivation can be much more powerful than the extrinsic motivation of praise, fame or money.[30]

And this kind of motivation is not the privilege of the eminent few. It can be shared by the obscure many. In the words of Csikszentmihalyi – who is best known for the concept of *flow*, a pleasurable state of mind that comes with being fully absorbed in an activity – 'when we are involved in [creativity], we feel that we are living more fully than during the rest of life.'[31] And when people are asked what they enjoy the most when they do things they like, they say 'designing or discovering something new'.[32]

And this is perhaps the most important message to creators from a world replete with sleeping beauties and names writ in water: the time to create unrecognised gems is well spent if creating itself comes from a place of pleasure, joy and love.

Even creators who despair at their lack of success may occasionally glimpse this truth. Among them is John Keats, who escaped this despair through long walks in nature, where he observed life thriving all around him without any thought of fame or fortune. These walks opened his mind to a realm where the causes of his suffering played little role, a realm that moved him deeply. It also left an imprint in some of his poems, where he glorifies 'fruit ripening in stillness' and 'the birth, life, death of unseen flowers'.[33]

ACKNOWLEDGEMENTS

ONLY A FEW OF MY collaborators and students are mentioned in the text by name, but this book would not have been possible without many others. Countless conversations with numerous young researchers who spent time in my research lab during the last ten years have shaped my thinking about the subject of this book. Among my colleagues I would like to thank Mar Albà, Manne Bylund, Jessica Cantlon, Peter Gärdenfors, Gene Hunt, Eric Libby, Marcelo Sánchez-Villagra and Carel van Schaik for providing valuable feedback on individual chapters. Wherever I did not follow their advice, the book may be the worse for it. I finished a first draft of the book during a research semester at the Stellenbosch Institute for Advanced Study in South Africa (STIAS), which provided a fantastically supportive and intellectually stimulating work environment. Without the help of STIAS and its devoted staff, this book would have taken me much longer to finish. My thanks also go to the numerous colleagues and fellow visitors that I have encountered over the years at the Santa Fe Institute. While I spend less time there than I wish I could, the Institute has remained a wellspring of stimulation, ideas and examples for this book. The University of Zürich and my colleagues there deserve thanks for helping create the supportive institutional environment in which projects like this can thrive.

My agent Lisa Adams helped me navigate the treacherous waters of the publishing industry and handled all contractual matters with aplomb. Furthermore, I am indebted to Sam Carter from Oneworld for his incisive comments on the manuscript, for numerous useful suggestions, and for his professionalism as an editor. Thanks also go to the rest of the editorial team at Oneworld, including Tom Feltham, Holly Knox, Paul Nash and Rida Vaquas, for their professionalism, promptness and thoroughness in getting this manuscript in shape for publication. Last but not least, I would like to thank my family for putting up with my absent-minded, cantankerous and impatient self while I finish complex projects like this.

NOTES

1. Dillon and Millay (1936).

INTRODUCTION

1. Termites invented this kind of farming independently from ants. For this and other examples of multiple discoveries in evolution see also Losos (2017).
2. See Arnold (2018).
3. See Tyndall (1863), Blüh (1952), as well as p. 318 of Merton (1961).

1: INSTANT INNOVATION

1. Agrawal and Konno (2009).
2. Farrell et al. (1991).
3. When it was coined in the mid-twentieth century, the notion of a key innovation referred to traits that allow an organism to interact in novel ways with its environment, but today the term increasingly refers to traits that promote

adaptive radiation. More specifically, a key innovation must be both necessary and sufficient for an adaptive radiation. See Heard and Hauser (1995), as well as Erwin (1992). This definition is useful in practice, because it allows one to identify key innovations by comparing the number of species between clades (groups of organisms with a single common ancestor) that have an innovation and others that do not. The examples in this chapter meet this criterion. However, the definition also has limitations. First, most clades differ in multiple traits and it may be difficult to identify which (if any) one is the key innovation. Second, some innovations may be useful to an organism, yet they may not lead to adaptive radiation. Third, some radiations may create species that are very similar to one another, with limited ecological diversity that indicates different lifestyles built upon a key innovation. (Many researchers distinguish between the number of species in a clade and their morphological or ecological diversity, often referred to as disparity.)

4. See Farrell et al. (1991). More precisely, the study used latex and resin canals as proxies for latex and resin production. I also note that here and elsewhere I use for simplicity the word 'family' in the colloquial, not the taxonomic sense. Any one such family could be a taxonomic family, but also a genus, order, class or any other standard taxonomic unit.

5. Agrawal and Konno (2009).

6. Mitter et al. (1988). Following these authors, I refer to insects that feed on the living tissues or sap of higher (vascular) plants, but not to insects that feed on planktonic algae or nectar.

7. More precisely, all this is referring to the structure of the upper molars. See Hunter and Jernvall (1995).

8. Uauy et al. (1995).

9. Hughes and Eastwood (2006).

10. Similar radiations occurred in Lake Tanganyika and Lake Victoria. See Genner et al. (2007), Kocher (2004), as well as Brawand et al. (2014). Some scientists have proposed that a cichlid-specific trait called the pharyngeal jaw – a secondary jaw in the pharynx or throat – is a key innovation; see Liem (1973). This trait may be necessary but is not sufficient for cichlid radiations: all cichlids share the pharyngeal jaw but not all of them radiated when colonising a lake; see Seehausen (2006). What is more, pharyngeal jaws may even accelerate species extinctions. See McGee et al. (2015).

11. Kocher (2004).

12. Maan and Sefc (2013).

13. MacFadden (2005).

14. See Geist et al. (2014), as well as p. 220 of Whittaker and Fernandez-Palacios (2007) for relevant dating information. An archipelago may be older than its oldest island today, because some islands may become submerged again after they have risen from the ocean. However, molecular clock dating shows that on Hawaii few organismal lineages are older than ten million years.

15. See McKinnon and Rundle (2002), as well as Bell and Aguirre (2013).

16. See Gibson (2005) and Colosimo et al. (2005).

17. See Bell et al. (2006) and Hunt et al. (2008). I note that these studies did not focus on the original colonisation of the lake, but the replacement of a low-armour species by an invading high-armour species, which subsequently lost its armour.

18. See Giles (1983).

19. See Bell et al. (2006) and Hunt et al. (2008). Stickleback evolution can proceed even faster than demonstrated in this paleontological study, i.e., on the time scale of decades in contemporary lakes. See Bell and Aguirre (2013) as well as Bell et al. (2004).

20. The data I cite is from Hunt (2007). More recent data can be found in Hunt et al. (2015) as well as Voje et al. (2018). See also Estes and Arnold (2007). Prolonged stasis can have several causes that can be difficult to distinguish in practice. See Hunt and Rabosky (2014), as well as Hansen and Houle (2004). Periods of stasis interspersed with short bursts of rapid change are reminiscent of the mode of evolution that palaeontologist Stephen Jay Gould has called punctuated equilibria (see Gould and Eldredge, 1977). Indeed, they are consistent with Gould's postulate that most evolutionary change is rapid and that slow gradual change is relatively rare. However, they make no statement about a second essential tenet of punctuated equilibria, namely that rapid change is usually associated with speciation.

21. See Seehausen (2006). One possibility is that recombination among genetically diverse cichlids is needed to create the kind of variation that fuelled cichlid radiations. See Meier et al. (2017).

22. Losos and Mahler (2010), as well as Stroud and Losos (2016) review relevant literature. One general pattern is that the overall outcome of repeated radiations is most similar when the radiations started from similar species.

23. See Bornman et al. (1973).

24. See Blount et al. (2018).

25. Gould famously argued that replaying life's tape would not lead to a flora and fauna resembling the one we know (Gould, 1990). Gould's speculation is hard to disprove, but evidence against it has been mounting in various forms. Among them are the multiple origins of innovations like hypocones, latex and many others we will hear about later. See Losos (2017) for a recent exploration of this issue.

26. The kind of experimental evolution I describe here starts from a population of genetically identical individuals. Other

kinds start from individuals that are not genetically identical, which has the advantage that the genetic variation which natural selection needs is not created during the experiment, but already exists at its beginning. In consequence, evolution can proceed even faster. However, a genetically variable starting population also has a disadvantage: it can make the DNA mutations that cause evolutionary change difficult to identify.

27. Graves Jr et al. (2017) and Phillips et al. (2016).
28. Lenski (2017) and 'Miniature Worlds' Anonymous (2020).
29. Dhar et al. (2013), Dhar et al. (2011), Sprouffske et al. (2018), Karve and Wagner (2021), as well as Karve and Wagner (2022).
30. See Tenaillon et al. (2012) for an example, and Blount et al. (2018) for a review, including some prominent exceptions. The latter article focuses on the role of contingency, i.e., the influence that different genetic starting points of evolution can have on the outcome of evolution, a related and fascinating topic.
31. Blount et al. (2008).
32. Blount et al. (2018), Blount et al. (2008), as well as Leon et al. (2018). Citrate utilisation actually required more than one mutation, where later mutations are contingent on previous ones.
33. Hall (1982).
34. For a survey of multiple antibiotic resistance mechanisms, see Blair et al. (2015). More generally, individuals in different replicate populations at the end of an evolution experiment often experience some mutations in the same genes, but also multiple mutations in different genes. Some of these mutations may be unique to a single replicate. See Lang et al. (2013), Tenaillon et al. (2012), as well as Sprouffske et al. (2018) for examples. One experimental challenge is to distinguish beneficial mutations from mutations that just 'hitchhike' with other mutations to spread through a population.

2: THE LONG FUSE

1. See Turnbull (2003) and Frazier (2005).
2. Frank et al. (1998).
3. Chernykh (2008).
4. See Piperno and Sues (2005), as well as Prasad et al. (2005).
5. Stromberg (2005).
6. See, for example, Ossowski et al. (2010). I am glossing over multiple details here, because they do not affect the general principle. First, not just neutral mutations, but also mutations that are slightly detrimental can be inherited, because natural selection is not perfect at eliminating such mutations. Second, all organisms occur in populations, and any one mutation takes place in only one organism of a population. This means that, on the one hand, it is not guaranteed that the mutation sweeps through the population and becomes part of a species' genetic heritage. On the other hand, it creates more opportunities for the ticking of the clock. For neutral mutations, however, it turns out that the number of individuals in a population is immaterial for the rate at which the clock advances (see Lynch, 2007). Third, any one clock can vary in the rate at which it ticks over time. Fourth, to infer an absolute time scale at which a molecular clock ticks, the clock needs to be calibrated. Fossil species whose position on the tree of life are well-known, and whose age has been measured through radiometric dating methods are important for such calibration (see Ho and Duchêne, 2014).
7. More precisely, both proteins contain not just one but multiple amino acid chains, each encoded by one gene. The gene atpB encodes one of multiple amino acid chains of atp synthase, and the gene rbcL encodes one of two amino acid chains in RuBisCO. Both of them are widely used to

reconstruct evolutionary relationships among plants, including grasses. See Bremer (2002).

8. Bar-On et al. (2018).

9. One complication is that clock speeds can vary on different branches of the tree of life, and corrections are necessary to account for such variation. Also, if a clock has ticked many times along one branch, chances are that the same letter may have been hit more than once by mutation. One can correct for such so-called multiple substitutions. See Felsenstein (2004).

10. More precisely, all the leaves of one branch correspond to what biological systematics calls a monophyletic group of species, which comprises all (and not just some) extant descendants of a common ancestor. This definition also includes all living species as a monophyletic group, because they share a common ancestor at the origin of extant life – the 'branch' that harbours them is the entire tree of life.

11. Bremer (2002).

12. Stromberg (2005).

13. I note that not all of these innovations originated at the same time, and some arose long before grasses became ecologically dominant. Also, not all of them are unique to grasses, nor are they all shared by all grasses. For example, only about half of all grasses have C_4 photosynthesis. See Sage (2004).

14. These and other grass traits are discussed in Coughenour (1985).

15. Sage (2004).

16. In addition to this novel biochemistry, many C_4 plants have also evolved a distinct leaf anatomy that helps them concentrate carbon dioxide around RuBisCO. I note that the advantage of C_4 plants is not universal, and C_3 plants may do better in some environments, such as cold environments, and at times when atmospheric CO_2 is high. About 50% of grasses are C_4 grasses. See Edwards and Smith (2010).

17. See Christin et al. (2008), as well as Edwards and Smith (2010).
18. Not all C_4 plants are grasses, and not all grasses are C_4 plants. Some grasses, like rice, have C_3 photosynthesis. C_4 photosynthesis originated up to 45 times independently in plants overall. As one would expect from independent innovations, C_4 plants are not identical in all details, but differ in their biochemistry and leaf anatomy. See Sage (2004).
19. Christin et al. (2008).
20. Arakaki et al. (2011).
21. Nishiwaki (1950).
22. Pickrell (2019). For an attempt to reconstruct some traits of the most recent common ancestor of mammals see Werneburg et al. (2016).
23. Luo (2007). To be precise, I am referring to eutherian mammals, which are those extant or extinct mammals most closely related to placental mammals like us, as opposed to marsupials like the kangaroo, or monotremes like the duck-billed platypus.
24. Ji et al. (2002).
25. Luo (2007).
26. Ji et al. (2002).
27. Luo et al. (2011).
28. Ji et al. (2006), Luo (2007).
29. Meng et al. (2006).
30. Luo and Wible (2005).
31. See Grossnickle et al. (2019). Yet another factor is suggested by Brocklehurst et al. (2021).
32. Xu et al. (2014).
33. See Xu and Norell (2004). Like many other details of bird evolution, it is somewhat contentious whether Archaeopteryx itself was warm-blooded, but bone growth patterns argue against it. See Erickson et al. (2009).

34. Norell et al. (1995).
35. Xu et al. (2014).
36. Although *Microraptor* was a small predator, it was large enough to feast on members of the other emerging group, the mammals. We know because the guts of some *Microraptor* fossils contain tree-dwelling creatures that resembled *Eomaia*, evidence that dinosaurs were not just competing with but also preying on mammals. See Larsson et al. (2010).
37. Field et al. (2018). That holds even if one takes into consideration that the hollow bones of birds do not fossilise well in general.
38. See Field et al. (2018), Claramunt and Cracraft (2015), as well as Feduccia (2003).
39. Field et al. (2018).
40. Heard and Hauser (1995).
41. See Jablonski (2017), as well as Kröger and Penny (2020).
42. I note that such innovations must at least survive (and thus succeed) long enough to become preserved as fossils, perhaps because they ensured survival in a narrow ecological niche. There may be countless others that did not even last that long.
43. Moreau et al. (2006).
44. Grimaldi and Agosti (2000).
45. Thatje et al. (2008) and Near et al. (2012). The ancestors of notothenioids were benthic fish – they lived on the ocean floor – but then radiated to occupy other ocean habitats. In addition to their antifreeze proteins, notothenioids also show other adaptations, such as a reduced density of their skeleton, which compensates for their lack of a swim bladder. These adaptations too originated long before the notothenioids eventually radiated. See Daane et al. (2019).
46. Stanley (2014) and van der Heide et al. (2012).
47. Plants also experienced delayed success, but we know less about the environmental triggers involved. See Knoll (2011).

48. See Erwin et al. (2011).
49. See chapter 4 of Erwin and Valentine (2013).
50. See p. 80 of Erwin and Valentine (2013), as well as Srivastava et al. (2008).
51. The beginning of the Cambrian explosion is often placed at 541 million years ago, but is subject to debate. Also, it is an explosion only compared to what came before. It unfolded over tens of million years, and may have involved more than one wave of innovation. See Wood et al. (2019).
52. See chapter 6 of Erwin and Valentine (2013) for a detailed discussion of my examples and multiple others from the bizarre Cambrian explosion fauna.
53. Erwin and Valentine (2013), p. 208.
54. Ibid., pp. 194–5.
55. Whittington (1975).
56. See Paterson et al. (2011).
57. See Cole et al. (2020), Gramling (2014), Wood and Erwin (2018), Fox (2016), Judson (2017), Reinhard et al. (2016), Wood et al. (2015), Sahoo et al. (2016), Sperling et al. (2015), Zhang et al. (2016), as well as Mills and Canfield (2014). One source of controversy is that planetary oxygenation has to be inferred from indirect evidence, such as the scarcity of easily oxidised minerals like pyrite. Another is that oxygenation has progressed not continually but haltingly, with oxygen concentrations fluctuating over time, because not only photosynthesis but also other, geochemical processes contribute to oxygenation. Thirdly, oxygenation was not uniform in space either. For example, the deep ocean is oxygenated less easily than surface waters, where most photosynthetic algae live.
58. Judson (2017). More precisely, molecular oxygen is a great terminal electron acceptor, to which electrons get transferred during the process of aerobic respiration.
59. Fenchel and Finlay (1995), as well as Catling et al. (2005).

60. Catling et al. (2005).

61. Tellingly, palaeontologists define the onset of the Cambrian itself not by the fossil of an animal's body but by a trace fossil called *Treptichnus pedum*, one of the first known vertical burrows. See p. 16 of Erwin and Valentine (2013).

62. See chapter 6 of Fenchel and Finlay (1995).

63. Shelled fossils are known from earlier times, but these are microfossils of single-celled organisms that protected against simpler, single-celled predators. See Porter (2011).

64. See Cole et al. (2020) for a discussion of oxygen and its alternatives.

65. See Bourke (2014).

66. Knoll (2011). I am referring to multicellularity in eukaryotes. It has originated multiple additional times in prokaryotes, which show simpler forms of multicellularity, such as biofilms. See Lyons and Kolter (2015).

67. Boraas et al. (1998).

68. In later cycles of the experiment, they used a centrifuge to accelerate settling. See Ratcliff et al. (2012).

69. Similar experiments are reported in Koschwanez et al. (2013), Ratcliff et al. (2013) and Ratcliff et al. (2015).

70. Powner et al. (2009), Xu et al. (2020), Miller (1953) and Miller (1998).

71. Schmitt-Kopplin et al. (2010). See http://www.lpi.usra.edu/meteor/metbull.php?code=16875 for documentation of the comet's impact.

72. Deamer (1998).

73. See Blain and Szostak (2014), as well as Szostak (2017). Fatty acids are only the simplest membrane building blocks. Today's cells use more complex phospholipids.

74. Sleep et al. (2001) and Chang (2008).

75. Rosing (1999), Pecoits et al. (2015), Bell et al. (2015), Nutman et al. (2016) and Dodd et al. (2017).

3: MOLECULAR MOTORWAYS

1. See Copley et al. (2012), Nohynek et al. (1996), Dai and Copley (2004), as well as Chang and Su (2003).
2. Bushman (2002).
3. Archer et al. (2011).
4. See Dai and Copley (2004), as well as Hlouchova et al. (2012).
5. Given pervasive horizontal gene transfer, it can be difficult to determine a bacterium's 'core' genome, the part of its genome that has never been horizontally transferred.
6. Boyle et al. (1992).
7. Grunwald (2002) and Neslen (2017).
8. Boyle et al. (1992).
9. See Bedard et al. (1987), Denef et al. (2004), Leigh et al. (2006), as well as Boyle et al. (1992).
10. Muller et al. (2003).
11. Stokstad (2007) and Purnomo et al. (2011).
12. Martin and Drijfhout (2009).
13. Regnier and Law (1968).
14. Scholz et al. (2015).
15. In the eukaryotic cells of our bodies, mitochondria are responsible for respiration, but respiration is much older than mitochondria, which evolved from a respiring prokaryotic ancestor. See Wang et al. (2011), as well as Kim et al. (2012). The question whether mitochondria have been central to eukaryotic success, and whether they represent an early or a later step in the evolution of eukaryotes, is controversial. See Lane and Martin (2010), Lynch and Marinov (2015), as well as Zachar and Szathmary (2017). Uncontroversial is that mitochondria eventually became central to eukaryotic energy metabolism.
16. I emphasise carbon, because it is such a central chemical element, but E. coli also needs other chemical elements, such

as nitrogen, where the same holds, i.e., *E. coli* can survive on multiple different sources of nitrogen. It is also remarkable that *E. coli* can survive in a minimal chemical environment that contains only a handful of different molecules, one each for each abundant chemical element, and some trace elements.

17. This does not mean that our metabolism catalyses fewer reactions. See Duarte et al. (2007). For example, our bodies produce entire classes of complex molecules that are central for cell communication in multicellular organisms, such as sterols and steroid hormones, which are less important or absent in bacteria. See Desmond and Gribaldo (2009).

18. For a detailed account, including a comparison between different *E. coli* strains, see Table S9 from Archer et al. (2011).

19. See Feist et al. (2007). The relationship between reactions and genes is not always one-to-one. Some reactions are catalysed by enzymes formed from multiple proteins, and some enzymes catalyse multiple reactions, as we shall also see in chapter 4.

20. See Feist et al. (2007), Segre et al. (2002), as well as Hartleb et al. (2016).

21. Kanehisa et al. (2016).

22. See Rodrigues and Wagner (2009), as well as Samal et al. (2010). The algorithm relies on a method called Markov Chain Monte Carlo sampling that modifies an existing metabolism like that of *E. coli* by adding or removing reactions until its reaction complement has been effectively randomised, while leaving its metabolic abilities unchanged. The method is popular because it allows effective randomisation of a complex metabolic system at reasonable cost of computation time.

23. Samal et al. (2010).

24. For examples see Koehn et al. (2009), Olszewski et al. (2010), Berg et al. (2007), as well as Hiratsuka et al. (2008).

25. Barve and Wagner (2013).
26. The same holds for metabolisms viable on primary carbon sources different from glucose. They also come in very different forms, and they are gratuitously viable on multiple further carbon sources.
27. See Meijnen et al. (2008). The authors did not report an explanation, but speculated that enzyme promiscuity, another cause of gratuitous innovation I discuss in chapter 4, may be responsible. Since experiments like these do not usually alter metabolism dramatically but modify enzymes and their regulation, that is a plausible explanation. I mention the example here to illustrate that metabolic innovations can arise even on the short time scales of laboratory evolution, and without conveying an immediate benefit.
28. Unfortunately, the road analogy is imperfect, because sometimes new food molecules that emerge are not part of a newly added pathway but far away from it. It's as if a new road through one neighbourhood could directly connect to other, far away neighbourhoods. We don't see that in daily life, but in metabolism it is commonplace. That's because a chemical network is more interconnected than any road network, such that a single new reaction can connect previously distant parts of metabolism. Mathematically speaking, a metabolic network does not exist on a two-dimensional surface like a road network, but in a higher-dimensional space.
29. See Barve and Wagner (2013).
30. Complexity of a different kind results in a similar innovation benefit. It comes from metabolisms designed to sustain life on not just one but multiple foods: they thrive on even more foods. Where the average metabolism designed to use one food source is actually viable on five others, metabolisms designed to use ten foods are actually viable on twenty-five others – that's fifteen innovations for free. See Barve and Wagner (2013).

31. I do not mean to imply that the consortium or its members are actively pursuing that goal. Rather, such efficiency is an outcome of ecological processes in which less efficient consortia are replaced by more efficient ones through the extinction of existing consortium members, and the successful invasion of new ones. See Wagner (2022).

32. Consortia do not always process food as in the relay race analogy. For example, one organism may create multiple by-products, each of them serving others as food for another consortium member. The thermodynamics of energy metabolism are also important. A given food alone may not yield enough energy for one organism, but when its waste products are removed by a second one that metabolises them, this yield can become positive. See Morris et al. (2013).

33. Bacterial multicellularity can take many forms, and labour sharing is only one of its benefits. See Lyons and Kolter (2015). For the origins of cyanobacterial multicellularity, see Schirrmeister et al. (2011) and Schirrmeister et al. (2013).

34. Stal (2015). More precisely, the nitrogenase-operating cells still harvest light, but they abandon the part of photosynthesis that creates oxygen, which leaves them dependent on the electrons, or so-called reducing equivalents, that would otherwise be produced by photosynthesis, and that they need to import from photosynthesising cells.

35. Libby et al. (2019).

4: GOOD VIBRATIONS

1. Clemente et al. (2015).
2. The non-governmental organisation Survival International dedicates itself to preserving the autonomy of this and other

tribal peoples. See https://www.survivalinternational.org/tribes/yanomami.

3. See http://www.proyanomami.org.br/frame1/ingles/saude.htm.

4. Sender et al. (2016). Although they are very abundant, the often cited estimate that the cells of our microbiome may outnumber our own body's cells by ten to one may be wrong.

5. Crofts et al. (2017) and Robinson et al. (2005). The same holds for its synthetic predecessor, the sulphonamides, i.e., resistance against them also evolved rapidly. See Davies and Davies (2010).

6. Murray et al. (2022).

7. Many of these genes confer antibiotic resistance when they are overexpressed in the laboratory, which is a hallmark of the promiscuity that is central to this chapter.

8. Clemente et al. (2015).

9. Yamaguchi et al. (2012).

10. Brown and Balkwill (2009).

11. Toth et al. (2010).

12. D'Costa et al. (2011). Although some descendants of these bacterial communities may be alive today, most bacteria from the environment cannot be cultured in the laboratory, such that it is easier to identify their resistance traits from their DNA, by expressing the proteins encoded in this DNA in other organisms in the laboratory.

13. Bhullar et al. (2012).

14. It bears mentioning that sulphonamides, a group of synthetic drugs, were widely used before penicillin.

15. Other natural roles for antibiotics have been proposed, for example that they can help cells communicate because they tend to occur in low, sub-lethal concentrations in the environment. However, they can inhibit competitors even at sub-lethal concentrations. See Yim 2007 and Abrudan 2015.

16. Carroll (1871).
17. Taylor and Radic (1994), as well as Quinn (1987).
18. See Thatcher et al. (1998).
19. One distinguishes between substrate promiscuity, the ability to catalyse the same kind of reaction but with different molecules (substrates), and catalytic promiscuity, the ability to catalyse different kinds of chemical reactions. See Copley (2017).
20. Biochemists also distinguish between enzymes whose biological function it is to catalyse a broad spectrum of chemical reactions, and promiscuous enzymes, where at least some of the reactions they catalyse have no physiological role. See Khersonsky and Tawfik (2010). In practice, it may be difficult to determine whether any one reaction serves no biological purposes, but several criteria can be helpful, i.e., whether an enzyme catalyses the reaction at a much lower rate than other reactions, whether other enzymes in the same organism catalyse the same reaction at a much higher rate, and whether the reaction involves synthetic molecules, such as synthetic antibiotics, that an organism is not likely to encounter in the wild.
21. Bush and Jacoby (2010), Weikert and Wagner (2012), as well as Baier and Tokuriki (2014).
22. I am referring to one kind of efflux pump from a family known as the RND family. See Murakami et al. (2002). Their motion is also referred to as a peristaltic motion, in analogy to the motion of our gut. There are multiple other promiscuous efflux pumps, with different primary functions and evolutionary origins. See Du et al. (2018). In addition, there are further antibiotic resistance mechanisms that I do not discuss at all, for example, to render cells impermeable to antibiotics, or to modify the proteins that antibiotics target, thus rendering them immune to their action.

23. Several of these pumps are formed by more than one poly-peptide, and synthesising multiple pumps requires making multiple copies of each polypeptide. See Du et al. (2018).
24. See Soo et al. (2011). These enzymes – carbonic anhydrases – have multiple roles in bacteria, including to stabilise a cell's pH. They are also targeted by some drugs, including some sulphonamides. See Supuran and Capasso (2017).
25. Karve (2021), Karve and Wagner (2022a), as well as Karve and Wagner (2022b).
26. Blount (2015).
27. Nam et al. (2012).
28. Notebaart et al. (2014).
29. See Toll-Riera et al. (2016). To be precise, she performed many more experiments, because she performed four replicate experiments for each environment. The organism she worked with was not *E. coli* but *Pseudomonas aeruginosa*, another popular organism for laboratory evolution experiments.
30. On average. The experiment had been extremely laborious as it was, so Macarena did not study the DNA mutations and the enzymatic changes they caused in great detail.
31. Huang et al. (2015). More precisely, the enzymes in question belong to the haloalkanoate dehalogenase superfamily, most of which transfer phosphates.
32. Dantas et al. (2008).
33. The consequences of bacteria's talent for horizontal gene transfer was on display in the example of pentachlorophenol from chapter 3, a man-made toxin that, unlike antibiotics, was not designed to kill bacteria.
34. This description fits regulators called activators, because they turn genes on. Other regulators are called repressors. They turn genes off by preventing their transcription.
35. Hall et al. (2004), as well as Gancedo and Flores (2008).
36. See Wistow and Piatigorsky (1988), Piatigorsky and Wistow (1989), as well as Piatigorsky (1998).

37. Some crystallins derive from proteins with other functions, such as to protect against heat. See Wistow and Piatigorsky (1988).
38. Mithofer and Boland (2012).
39. Dinan (2001), as well as Mithofer and Boland (2012).
40. Mithofer and Boland (2012), Huang et al. (2016), as well as Steppuhn et al. (2004).
41. Mithofer and Boland (2012).
42. See figure 4 of Weng and Noel (2012). For promiscuity in plants see Westfall et al. (2012), O'Maille et al. (2008), Schuler and Werck-Reichhart (2003), as well as Weng et al. (2012).
43. See Gibbs and Hosea (2003), as well as Rendic and DiCarlo (1997).
44. See Schuler and Werck-Reichhart (2003), as well as Grebenok et al. (1996). Cytochromes P450 also occur in many animals, including humans, where they help biosynthesise molecules like hormones, but also detoxify drugs and other toxins. They are often thought of as enzymes with broad specificity rather than promiscuity, because their broad specificity may be important for what they do, but even so, many of their molecular products may not be biologically important in all environments.
45. Richardson and Pyšek (2006) discuss the multiple factors that influence the invasiveness of a species and the invasibility of a habitat.
46. Finch (2015).
47. The ecological and economic impact of water hyacinths, as well as efforts to control them are well described in Theuri (2013). As a consequence of being released from natural enemies in their new habitat, invasive plants may need to spend less energy on defense chemicals, which allows them to reproduce faster. See Inderjit (2012).
48. Pinzone et al. (2018) and Raguso (2008).

49. Richardson et al. (1994).
50. Kimura et al. (2015).
51. Chen et al. (2017).
52. Chen et al. (2017).
53. See p. 175 of Darwin (1872).
54. Except that fish control buoyancy by excreting gas into the swim bladder, thus changing its volume, whereas submarines fill ballast tanks to varying degrees with sea water.
55. Luo (2007).
56. See p. 454 of Futuyma (1998).
57. Gould and Vrba (1982).
58. Weaver et al. (1985), Gould and Vrba (1982), as well as Gerhart and Kirschner (1998).
59. True and Carroll (2002).
60. Gould and Vrba (1982) also called such useless traits non-aptations, but the term never caught on.
61. D'Ari and Casadesus (1998).

5: THE BIRTH OF GENES

1. Hoyle (1950), p. 19.
2. Jacob (1977).
3. See Lynch and Conery (2000).
4. In practice, additional factors often play a role. For example, mutations that reduce the amount of a translated protein can be harmful. See Cook et al. (1998).
5. This is especially evident from whole genome duplications, where all genes in a genome are duplicated. See Kellis et al. (2004).
6. The term 'selfish gene' was coined by biologist Richard Dawkins in an eponymous book that applied the term to all genes. It is often used more narrowly, like I do here, for

genes that promote their own spreading without helping their host organism to survive.

7. Lynch and Conery (2000).

8. Mizutani and Sato (2011), as well as Schuler and Werck-Reichhart (2003).

9. And he took inspiration from earlier work, most notably that of Ohno (1970).

10. For an excellent account of how the workings and structure of the ribosome were discovered, see Ramakrishnan (2019).

11. For simplicity, I am omitting other requirements for a gene's DNA sequence. These include a binding site on DNA for polymerase itself, DNA words that signal the polymerase to stop transcribing and so on. Most of what I say here applies to eukaryotes, organisms whose cells, like ours, contain a nucleus, where most *de novo* genes have been discovered. The genes of prokaryotes vary in their sequence require-ments. For example, they require a specific sequence on the transcript called a ribosome binding site that is necessary for translation to start.

12. O'Bleness et al. (2012).

13. Rogers and Gibbs (2014).

14. Yang et al. (2018).

15. Khalturin et al. (2009). A more precise term is a taxonomi-cally restricted gene, which refers to genes that occur only in a limited number of closely related species.

16. Carvunis et al. (2012) and Khalturin et al. (2009).

17. The alternative is that such genes are in fact old but fast-evolving genes whose DNA sequence has diverged be-yond recognition. Approaches to identify new genes that rely on sequence similarity alone are controversial, because they can have difficulty distinguishing between the two al-ternatives. See Moyers and Zhang (2018), Casola (2018), as well as Domazet-Loso et al. (2017). This is why I here focus

on studies that use additional criteria, such as a genomic region's transcription that is unique to a species.

18. Comparing genomes is necessary but not sufficient for such a discovery. One also needs to compare the species' transcriptomes, the complete set of RNAs transcribed from their genomic DNA.

19. See Levine et al. (2006) and Zhao et al. (2014).

20. See Chen et al. (2010). Not all of the new genes in this study originated from scratch, but some did. Also, engineering techniques to reduce transcription are usually not perfect. In the jargon of the field, they knock down – reduce – transcription rather than knocking it out. It is even more remarkable that a mere reduction of a young gene's transcription can be lethal.

21. See Halligan and Keightley (2006).

22. Random means that their letter sequence does not resemble any other gene, and that it is not constrained in its evolution by the negative consequences of mutations. It does not mean that all nucleotide letters are equally likely to occur in such a sequence, because mutations do not create all nucleotides with equal probability.

23. See Yona et al. (2018). These experiments were conducted in *E. coli*, and aimed to create a promoter, a stretch of DNA that can be recognised by RNA polymerase itself, and not by a regulator of transcription. This simplification does not affect the central principle, that random DNA is likely to harbour signals for transcription. I also note that the DNA signals that help trigger transcription in many eukaryotes are shorter than those *E. coli* uses, meaning that they can occur even more easily by chance alone.

24. Adams et al. (2000).

25. Zhao et al. (2014).

26. Kellis et al. (2003).

27. Carvunis et al. (2012).
28. Ladoukakis et al. (2011), as well as McLysaght and Guerzoni (2015). The latter number refers to ORFs longer than 33 nucleotides. There may be many more shorter ORFs.
29. Reinhardt et al. (2013), Cai et al. (2008), Xiao et al. (2009) and Baalsrud et al. (2018). I mentioned antifreeze proteins already in previous chapters. Some of them evolved *de novo*, whereas others evolved from pre-existing proteins.
30. Tripathi et al. (1998), Stepanov and Fox (2007), as well as Knopp et al. (2019).
31. Carvunis et al. (2012) and Schmitz et al. (2018).
32. Tompa (2012). It is controversial whether new proteins are more or less disordered than old proteins. See Wilson et al. (2017), Casola (2018) and Yu et al. (2016).
33. Andrews and Rothnagel (2014).
34. For an example from human sequence evolution see Ruiz-Orera et al. (2015).
35. Carvunis et al. (2012).
36. Carvunis et al. (2012) and Schmitz et al. (2018).
37. I am referring to so-called long non-coding RNA, commonly defined as RNA with a length greater than 200 base pairs. See Iyer et al. (2015) and Encode Project Consortium (2012). Such pervasive transcription does not amount to a huge waste of resources, because most such DNA is transcribed much more rarely than other, more important DNA, amounting to a small fraction of total transcription and energy expenditure. See Ruiz-Orera et al. (2015), as well as Neme and Tautz (2016).
38. See Meader et al. (2010) and Graur et al. (2013). The proportion of neutrally evolving DNA varies widely among eukaryotic species, but this proportion is generally substantial. For an example from yeast see Carvunis et al. (2012). Whether DNA evolves neutrally is usually assessed

by comparing its rate of evolution to regions of the genome that are neither expressed nor serve any other known function, such as ancestral repeat sequences. See Wiberg et al. (2015).

39. Based on information from six species as different as humans, mice and fish. See Ruiz-Orera et al. (2015). I am again referring to long non-coding RNA, a misnomer really, since a large fraction of such RNA may actually encode protein.

40. Ruiz-Orera et al. (2018).

41. Kutter et al. (2012). Strictly speaking, the mice in question are subspecies, i.e., *Mus musculus musculus* and *Mus musculus castaneus*. For related evidence from fruit flies see Palmieri et al. (2014).

42. See Neme and Tautz (2016).

43. Although most dead genes stay dead, and new genes arise elsewhere in the genome, dead genes can occasionally become resurrected through mutations. See Bekpen et al. (2009).

44. See Palmieri et al. (2014) for an example of this pattern from fruit fly evolution. See also Schmitz et al. (2018).

45. Studer et al. (2016).

46. Martins et al. (2007).

47. In fact, these genes do not even exist in other species within the genus *Daphnia*. This information is based on the genome of *Daphnia pulex* and its relationship to the closely related *Daphnia magna*. See Colbourne et al. (2011).

48. Ruiz-Orera et al. (2015).

6: THE CROW AND THE PITCHER

1. The examples in this paragraph and many others can be found in Shumaker et al. (2011). The dental floss example is

from Watanabe et al. (2007). I note that manufacturing tools is harder than just using them. I also note that the ingenuity of tool use is not always invented or learned by an individual. Sometimes it has been written into a species' genome by natural selection.

2. For more rigorous – and verbose – definitions of a tool see chapter 1 of Shumaker et al. (2011). I note that even more rigorous definitions leave borderline cases. For example, chimpanzees eat medicinal herbs to fight gut parasites, but is the active ingredient of such a herb a tool? See Huffman (2003). And killer whales whip up powerful waves to flush the seals they hunt from ice floes, but can the very medium in which you live be a tool?

3. See chapter 7 of Shumaker et al. (2011).

4. Scally (2016).

5. Whiten et al. (1999), Fujii et al. (2015) and van Schaik et al. (2003a).

6. Some researchers make a distinction between imitation, which is characteristic of human children and requires that the child can envision the goals and intent of the observed person, and emulation, which is more characteristic of chimpanzees, and mirrors how the observed animal changes the environment. See Tomasello (1999).

7. See p. 27 of Reader and Laland (2003).

8. Whiten et al. (2005).

9. An important difference is that human culture is highly cumulative, and relies on knowledge accumulating over generations, whereas the cumulative culture of animals, where it exists, may be simpler.

10. Falótico et al. (2019) and Haslam et al. (2016). Chimpanzees in West Africa may have been cracking nuts with stones for even longer. See Mercader et al. (2007).

11. Cited on p. xii of Shumaker et al. (2011).

12. van Schaik et al. (2003b).
13. McGrew et al. (1997).
14. See Boesch et al. (1994). In capuchin monkeys, two years have been reported. See Ottoni and Izar (2008).
15. Whiten et al. (2001).
16. Sargeant et al. (2007) and Mann et al. (2008).
17. Mann and Patterson (2013).
18. See Bacher et al. (2010), Mann and Patterson (2013) and Krützen et al. (2005).
19. See Kawai (1965). This and other examples of multiple discoveries are also discussed in Tomasello (1999).
20. Tomasello (1999) and Kawai (1965).
21. Hunt and Gray (2003).
22. Kenward et al. (2011). I note that this example is different from those that precede it, because the ant lion's behaviour is likely encoded in genes. However, the same principle applies. Co-option of an existing behaviour facilitates not just the discovery of a new behaviour by an individual, but also the evolution of such a behaviour in a species.
23. Thouless et al. (1989).
24. Tebbich et al. (2001). An important question in this context is whether genetic differences may cause behavioural differences between different groups of animals, and rigorous studies address this possibility. Tebbich et al. (2001) tried to minimise genetic differences between their two study groups – finches reared with and without a tool-use model – by assigning siblings from the same family to both groups.
25. Hunt and Gray (2004).
26. Taylor et al. (2007).
27. Wimpenny et al. (2009).
28. Haslam (2013).
29. See Haslam (2013) and Huffman et al. (2008).
30. Bird and Emery (2009).

31. Tebbich et al. (2002).
32. Genetic differences in the propensity for tool use between northern and southern populations are not likely to be responsible, partly because even some Alaskan populations behave more like Californian populations. See Fujii et al. (2015).
33. Ottoni and Izar (2008).
34. Spagnoletti et al. (2012).
35. Koops et al. (2013) and Koops et al. (2014).
36. For the role of opportunity in orangutans, see especially Koops et al. (2014), but also Fox et al. (2004), van Schaik et al. (2003b) and van Schaik et al. (2003a). For more general evidence favouring the opportunity over the scarcity hypothesis, see especially Sanz and Morgan (2013), Koops et al. (2013), as well as Spagnoletti et al. (2012). Exceptions do of course exist, such as perhaps my earlier example of Galapagos woodpecker finches in arid environments.

7: NUMBERING NEURONS

1. See Wynn (1992), Cantrell and Smith (2013) and chapter 2 of Dehaene (2011). One important question is whether subjects in such experiments count objects, or whether they estimate continuous quantities that will often be correlated with the number of objects, such as the total surface area of the number of objects. Carefully conducted studies control for such factors, for example by reducing the surface area per object as the number of objects increase. They reveal that subjects do not just quantify correlates of object number, but that such correlates influence number estimates. See, for example, Feigenson et al. (2002), Cantrell and Smith (2013), as well as Ferrigno et al. (2017).

2. We feel that subitising is instant, but our response time does in fact increase with the number of objects. See chapter 3 of Dehaene (2011).

3. The law does not apply to all sensory modalities, and even where it applies, it may not hold over the whole range of possible stimulus intensities.

4. Human adults typically have a so-called Weber fraction below 0.2, meaning that they can readily distinguish numbers of objects that differ by more than 20 percent. See Pica et al. (2004).

5. Halberda et al. (2008) and Feigenson et al. (2013).

6. One can even manipulate the accuracy of the approximate number system in children, which changes their performance on a subsequent maths test. See Wang et al. (2016).

7. Mazzocco et al. (2011).

8. See 'Clever Hans' Anonymous (1904) and 'A horse' Anonymous (1911).

9. See Piantadosi and Cantlon (2017), as well as Strandburg-Peshkin et al. (2015). Baboons do not use a potentially confounding continuous quantity, the total mass of the group, as a basis of their decision. Rigorous experiments can help eliminate such continuous quantities as the basis for a numerosity judgment. See Cantlon (2018). Unfortunately, such experiments are not possible in all animals and they are rarely possible in the wild.

10. Nieder (2018a).

11. Benson-Amram et al. (2018), Panteleeva et al. (2013), Nieder (2018a), Smirnova et al. (2000), Rose (2018), as well as Agrillo and Bisazza (2018).

12. Smart honeybees, I should add, because only a minority manage this task. See Skorupski et al. (2018).

13. For all we know, animals cannot represent the numerosity zero symbolically, but they grasp the numerosity zero. See Howard et al. (2018), Nieder (2016) and Nieder (2018b).
14. As quoted in Nieder (2016).
15. See Nieder (2018a).
16. See Skorupski et al. (2018), as well as Dehaene and Changeux (1993).
17. Nasr et al. (2019). For earlier efforts to study the 'number sense' of artificial neural networks see Dehaene and Changeux (1993), Verguts and Fias (2004), as well as Stoianov and Zorzi (2012).
18. Cited in Butterworth et al. (2018).
19. d'Errico et al. (2018).
20. The current, highly efficient place-position decimal system of encoding numbers arose even later. See chapter 4 of Dehaene (2011).
21. Tattersall (2009).
22. Few numerals are no principal obstacle to express arbitrarily large numbers. After all, the binary system used in digital electronics combines only the two numerals one and zero. However, many languages – forty-four percent of the Australian languages, for example – do not compose larger numbers out of numerals. See Bowern and Zentz (2012).
23. Epps et al. (2012), as well as Bowern and Zentz (2012).
24. Pica et al. (2004).
25. Dehaene and Cohen (2007).
26. Anderson (2010).
27. In people, such recordings are usually performed with volunteer patients who have to undergo neurosurgery for conditions such as epilepsy.
28. Others include the inferior temporal lobe and the prefrontal cortex. The intraparietal sulcus is especially well-studied.

See Amalric and Dehaene (2018), as well as Nieder and Dehaene (2009).

29. Nieder and Dehaene (2009), as well as Eger et al. (2003).

30. Amalric and Dehaene (2018).

31. Also, in both monkeys and humans, neurons in these regions are most sensitive to numerosity itself rather than to related quantities, such as to the total surface area of a number of dots. See Cantlon (2018) and Ferrigno et al. (2017). Continuous quantities such as surface area can, however, influence numerosity judgments. See Cantrell and Smith (2013).

32. Kutter et al. (2018), as well as Nieder and Dehaene (2009).

33. Diester and Nieder (2007), as well as Nieder and Dehaene (2009).

34. This is not to say that maths skills are completely independent of language. Mathematicians need words to communicate, and even the language of non-mathematicians in Western culture is replete with references to numbers. Also, it is not the case that the brain regions activated by mathematics are exclusively devoted to it. They, like many other regions in the cortex, are involved in more than one task, such as non-mathematical problem-solving and logical inference. See Amalric and Dehaene (2018).

35. See Nuñez (2017) and Nieder (2017) for the evolutionary origins of the number sense, and Hauser et al. (2002), as well as Pinker and Jackendoff (2005), for the evolution of language.

36. See Amalric and Dehaene (2016), Amalric and Dehaene (2018).

37. Puchner (2017).

38. See chapter 2 of Dehaene (2009).

39. Dehaene and Cohen (2011) call this region the 'visual word form area', a more technical term that is widely used.

40. See Dehaene et al. (2010), as well as Dehaene and Cohen (2011).

41. The region does not respond exclusively to letters and words, but also to other objects, just like many other cortical regions do not have sharp boundaries but overlap substantially with others. See chapter 2 of Dehaene (2009) and Vogel et al. (2014). Whether regions adjacent to it have evolved to recognise specific classes of objects or not is in itself a fascinating and debated question. See Arcaro and Livingstone (2021).

42. Although it is not clear yet whether reading incurs a cost in terms of the recognition of other objects. See chapter 5 of Dehaene (2009).

43. It is not necessarily individual neurons that suffice to recognise any one object with high precision, but entire ensembles of such neurons. See Rolls (2012), Rolls (2000), as well as chapter 3 of Dehaene (2009).

44. Rolls (2012) and Rolls (2000).

45. Chapter 3 of Dehaene (2009).

46. Changizi and Shimojo (2005), as well as Changizi et al. (2006).

47. Similar shape frequencies occur in computer-generated images of human buildings, and they also correlate highly with shape frequencies in written characters, showing that these shape frequencies are not just characteristics of the natural environment but of a broader class of objects in the world around us.

48. An important exception to this invariance regards mirror-symmetry in written characters. Because a character and its mirror image usually do not play the same role in a writing system, our brains should distinguish the two, but this distinction does not come naturally, precisely because of invariance. Learning this distinction thus poses special

challenges when children learn to read. See chapter 7 of Dehaene (2009).

49. See Dehaene and Cohen (2011) for application of Dehaene's cortical recycling principle to reading.

50. See pp. 150 and 139 of Dehaene (2009).

51. Whether the common ancestor of humans and chimps had a chimp-like hand or a human-like hand is still subject to debate. See Almécija et al. (2015). If its hand was more human-like, even fewer modifications may have been needed.

52. Diogo et al. (2012) and Tocheri et al. (2008).

8: CONCEALED RELATIONSHIPS

1. Quoted on p. 138 of Root-Bernstein and Root-Bernstein (1999).

2. See chapter 4 of Gamow (1966), as well as chapter 8 of Root-Bernstein and Root-Bernstein (1999).

3. Kilgour (1963).

4. See Gentner and Jeziorski (1989), Root-Bernstein and Root-Bernstein (1999), as well as chapter 5 of Pinker (2007).

5. The central importance of analogies for human cognition is underscored by the observation that children tacitly use analogies when learning abstract concepts like the size of a set of objects. See Gentner and Hoyos (2017), as well as Gentner (2010).

6. Cameron (2008).

7. Lakoff and Johnson (1980).

8. Lakoff (1993).

9. Lakoff (1993), as well as Lakoff and Johnson (1980).

10. Lakoff (1993).

11. Sometimes, the link between a metaphor and its origin ruptures completely. For example, most people do not

recognise the expression *a blockbuster movie* as metaphorical, because they do not know – as a previous generation did – that a blockbuster was a name for a bomb that could obliterate an entire city block. When that link to the source domain dies, the life cycle of a metaphor has reached its end. See Bowdle and Gentner (2005) for a detailed analysis of novel and conventional metaphors, as well as for the blockbuster example.

12. Hume (1740).
13. Talmy (1988) and chapter 4 of Pinker (2007).
14. Casasanto and Boroditsky (2008), Lakoff (1993) and chapters 4 and 5 of Pinker (2007).
15. Casasanto and Boroditsky (2008), as well as Boroditsky (2000).
16. Other experiments by the same authors show that estimates of duration affect estimates of distance to a much smaller extent. See Casasanto and Boroditsky (2008). I note that language can modulate the extent to which estimates of time can be affected by spatial quantities. See Bylund and Athanasopoulos (2017).
17. Chapter 67 of Kandel et al. (2013), and Moser et al. (2014).
18. Constantinescu et al. (2016). More generally, conceptual spaces may be crucial for abstract thinking, including metaphorical thinking. See Gärdenfors (2000).
19. Aronov et al. (2017).
20. Anderson (2010).

9: REINVENTING THE WHEEL

1. I am referring specifically to wheeled transport here and not to other uses of wheels. For example, the ancient Egyptians used potter's wheels long before they used wheeled transport. An overview of the wheel's history is given in Bulliet (2016).

2. Law (1980).
3. Wagner (2019), Wagner and Rosen (2014), Simonton (1999a), as well as Campbell (1960).
4. See Larson et al. (2014). An important difference is that agriculture in ants and termites is a product of biological evolution, whereas human agriculture is a product of cultural evolution. That difference, however, should not distract from the common principle of multiple discovery.
5. See p. 17 of Tully (2011), which reviews the history of rubber. At the time, huge rubber plantations in Southeast Asia that had been founded when seeds of the rubber tree *Hevea brasiliensis* had been smuggled out of Brazil in 1876 provided most of the world's rubber. When these plantations came under Japanese control, the Allied powers had a huge problem in supplying their industries with enough rubber. The problem was solved when synthetic rubber made of petroleum products was discovered.
6. Hosler et al. (1999).
7. Ogburn and Thomas (1922).
8. Merton (1961), as well as Ogburn and Thomas (1922).
9. Alfred (2009).
10. Cited in Lohr (2007).
11. Chapter 5 of Waller (2004) provides a concise overview of the history of scurvy.
12. Lamb (2001), p. 117.
13. Rajakumar (2001).
14. At least one similar, although perhaps unintended trial took place before Lind's time. In 1601, one Captain James Lancaster commanded four ships on an expedition to Sumatra, during which most sailors were sickened by scurvy and many died, except those on his own ship, which had received daily rations of lemon juice. He reported his observations to the Admiralty, but the cure was not widely adopted. See p. 117

of Waller (2004) and Mosteller (1981). If one considers 1601 as the pivotal date, then more than two centuries passed until citrus fruit were widely adopted as antiscorbutics.

15. See chapter 5 of Waller (2004). Just as during earlier centuries, individual seafarers did bet on the right antiscorbutics. Among them was Captain James Cook, whose 1768 South Sea expedition was not strangled by scurvy, because he periodically bunkered fresh fruit and vegetables.

16. Cited on p. 125 of Waller (2004).

17. See Cowan (1999) for relevant background material and references. Absorption and further refrigerator designs were also co-developed and patented by Albert Einstein and Leo Szilard, but even the endorsement of two such prominent physicists didn't ensure commercial success. See Dannen (1997). On the efficiency of absorption and compression refrigerators, see Bansal and Martin (2000).

18. Bell (2014).

19. Useful sources on the history of cardiac pacemaking include Aquilina (2006), Nelson (1993), Mulpuru et al. (2017) and Ward et al. (2013).

20. Altman (2002).

21. As cited on p. 247 of Ward et al. (2013).

22. Altman (2002) and Butler (2010).

23. Morton (2013), Mooney (2011), as well as Hausman and Kellogg (2015).

24. Morton (2013).

25. See p. 84 of Simonton (1988).

26. The term sleeping beauty for such papers goes back to Van Raan (2004). See Li and Shi (2016) for a paper's 'heartbeat', i.e., its citation rate.

27. Ke et al. (2015). The paper is not the first to use bibliometric analysis to identify sleeping beauties in the published literature, but its quantitative approach is especially compelling.

See also Redner (2005), Du and Wu (2018), El Aichouchi and Gorry (2018), Yeung and Ho (2018), Ohba and Nakao (2012), Van Calster (2012), Garfield (1990), Garfield (1989b) and Garfield (1989a).

28. That pattern notwithstanding, papers with very large beauty coefficients are rare. See Ke et al. (2015).

29. Darwin thought that a process then called pangenesis was responsible, which turned out to be wrong.

30. Galton (2009). It did not help that Mendel published his work in an obscure journal for local naturalists. See Mendel (1866).

31. Kottler (1979).

32. Garfield (1989b) and Balla et al. (2009). The original paper is Hokin and Hokin (1953). This is also an example of what has been called a premature discovery, where the rest of a research field needs to catch up to the discovery, so that it can be integrated into existing knowledge.

33. This example is discussed in Ohba and Nakao (2012), where the authors argue that the fact that the paper by Urayama (1971) was originally published in Japanese was not the cause of its neglect.

34. More precisely, these are more recent papers that cite the previously unrecognised paper, and also show a pattern of co-citation with it.

35. Ke et al. (2015).

36. Koestler (1964).

37. Kardos and Demain (2011). State-of-the-art sulfa drugs had serious side effects and were not effective against some infections. See Davenport (2012).

38. See Hardy (1967). As a pacifist who lived through two world wars, Hardy was especially concerned about how mathematics and science could be co-opted for waging war.

39. As quoted in Burr and Andrew (1992).

40. See Guterl (1994), as well as Schumayer and Hutchinson (2011).
41. For this and other examples, see Flexner (1939).
42. Table S4 of Ke et al. (2015), Van Calster (2012) and Bhanoo (2015).
43. See pp. 475–7 of Brockman (2017).

10: SLEEPING BEAUTIES

1. See p. 357 of Merton (1961).
2. Schwab (2012).
3. Ukuwela et al. (2013).
4. Denoeud et al. (2014).
5. Conway Morris (2003).
6. Gallagher (2015).
7. Losos (2017) and Blount et al. (2018). The constituent traits of such singletons, however, may not be so unique. The platypus is a bizarre animal, but its bill resembles that of a duck, the webbed feet resemble those of an otter, and the fur that of a sea otter. Similarities to body features of other organisms are enumerated on p. 328 of Losos (2017). In addition, it is not always easy to decide whether a new feature of an organism has been discovered multiple times. Is flight in insects convergent with flight in bats, even though the wings of these originated from completely different body structures? And what about flight in bats and pterosaurs – wings in both organisms originated from the arm, but from different parts of the arm? See chapter 3 of Losos (2017).
8. Lewis (1994).
9. Recent on the time-scales of evolution, that is, in the last few million years.

10. See Pritchard (1989). I am glossing over many aspects of radar's fascinating history, such as the post-war fight over priority within the British radar community, and parallel American efforts to develop radar.

11. See 'How was Hangul invented,' Anonymous (2013), as well as the National Institute of Korean Language at http://www.korean.go.kr/eng_hangeul/.

12. Chapter 5 of Hartl and Clark (2007), as well as chapter 14 of Barton et al. (2007).

13. More precisely, hemoglobin consists of four amino acid chains, two identical ones called α and another two identical ones called β, which are encoded by two separate genes, each of which has two copies. The sickle-cell mutation occurs in the gene encoding the β-chain.

14. Poolman and Galvani (2007).

15. Lynch and Walsh (1998).

16. Dean and Scharnhorst (1990).

17. McKay (1999). On a similar note, Bach's beloved cello suites were thought of as technical exercises until Pablo Casals started performing them in the early twentieth century. See Siblin (2010).

18. See https://neglectedbooks.com and https://www.nyrb.com.

19. Marche (2015).

20. Thompson (2017), p. 48.

21. Ibid., pp. 231–3.

22. Chapter 6 of Stross (2007).

23. Edison had competition, for example in companies that promoted electric arc lighting, which was much brighter than incandescent light bulbs. Edison won that fight but he lost others, for example that for direct current, which he thought was safer, but was more expensive to produce than alternating current. See chapters 6 and 8 of Stross (2007).

24. The incandescent light bulb is yet another one of those inventions with many parents. Although Edison was not the first to develop it, he developed it to commercial maturity, and this contribution arguably endured because he created the electrified environment in which his light bulb could thrive.
25. Caro (1982).
26. Albert (2019), p. 103.
27. Occasionally, some scientist claims otherwise, that evolution has foresight, but invariably, such claims have eventually been debunked. For example see the postulate of directed mutation by Cairns et al. (1988), and its resolution by Hendrickson et al. (2002). See also Lenski and Mittler (1993) and Wagner (2012).
28. Simonton (1988), p. 93, as well as Simonton (1994), Simonton (1999b), Sinatra et al. (2016) and Stern (1978).
29. See chapter 5 of Csikszentmihalyi (1996) for the preceding six quotes.
30. See Csikszentmihalyi (1996), as well as Amabile (1998) for a link between creativity and motivation.
31. Csikszentmihalyi (1996), p. 2.
32. Ibid., p. 108.
33. See Ward (1986), pp. 25–6.

BIBLIOGRAPHY

Adams, M.D., Celniker, S.E., Holt, R.A. et al. 2000. The genome sequence of *Drosophila melanogaster*. *Science* **287**, 2185.

Agrawal, A.A. and Konno, K. 2009. Latex: A model for understanding mechanisms, ecology, and evolution of plant defense against herbivory. *Annual Review of Ecology Evolution and Systematics* **40**, 311.

Agrillo, C. and Bisazza, A. 2018. Understanding the origin of number sense: A review of fish studies. *Philosophical Transactions of the Royal Society B: Biological Sciences* **373**, 20160511.

Albert, D. 2019. *Are We There Yet? The American Automobile Past, Present, and Driverless*. W.W. Norton & Company, New York, NY.

Alfred, R. 21 October 2009. Oct. 21, 1879: Edison gets the bright light right. *Wired Magazine*.

Almécija, S., Smaers, J.B. and Jungers, W.L. 2015. The evolution of human and ape hand proportions. *Nature Communications* **6**, 7717.

Altman, L.K. 18 January 2002. Arne H.W. Larsson, 86; had first internal pacemaker. *The New York Times*.

Amabile, T.M. 1 September 1998. How to kill creativity. *Harvard Business Review* **76**, 77–78.

Amalric, M. and Dehaene, S. 2016. Origins of the brain networks for advanced mathematics in expert mathematicians. *Proceedings of the National Academy of Sciences of the United States of America* **113**, 4909.

Amalric, M. and Dehaene, S. 2018. Cortical circuits for mathematical knowledge: Evidence for a major subdivision within the brain's semantic networks. *Philosophical Transactions of the Royal Society B: Biological Sciences* **373**, 20160515.

Anderson, M.L. 2010. Neural reuse: A fundamental organizational principle of the brain. *Behavioral and Brain Sciences* **33**, 245.

Andrews, S.J. and Rothnagel, J.A. 2014. Emerging evidence for functional peptides encoded by short open reading frames. *Nature Reviews Genetics* **15**, 193.

Anonymous. 2 October 1904. 'Clever Hans' again. *The New York Times*.

Anonymous. 23 July 1911. A horse – and the wise men. *The New York Times*.

Anonymous. 8 October 2013. How was Hangul invented? *Economist*.

Anonymous. 2020. Miniature worlds. *Nature Ecology & Evolution* **4**, 767.

Aquilina, O. 2006. A brief history of cardiac pacing. *Images in Paediatric Cardiology* **8**, 17.

Arakaki, M., Christin, P.A., Nyffeler, R. et al. 2011. Contemporaneous and recent radiations of the world's major succulent plant lineages. *Proceedings of the National Academy of Sciences of the United States of America* **108**, 8379.

Arcaro, M.J. and Livingstone, M.S. 2021. On the relationship between maps and domains in inferotemporal cortex. *Nature Reviews Neuroscience* **22**, 573.

Archer, C.T., Kim, J.F., Jeong, H. et al. 2011. The genome sequence of *E. coli* W (ATCC 9637): Comparative genome analysis and an improved genome-scale reconstruction of *E. coli. BMC Genomics* **12**.

Arnold, F.H. 2018. Directed evolution: Bringing new chemistry to life. *Angewandte Chemie International Edition* **57**, 4143.

Aronov, D., Nevers, R. and Tank, D.W. 2017. Mapping of a non-spatial dimension by the hippocampal–entorhinal circuit. *Nature* **543**, 719.

Baalsrud, H.T., Tørresen, O.K., Solbakken, M.H. et al. 2018. *De novo* gene evolution of antifreeze glycoproteins in codfishes revealed by whole genome sequence data. *Molecular Biology and Evolution* **35**, 593.

Bacher, K., Allen, S., Lindholm, A.K. et al. 2010. Genes or culture: Are mitochondrial genes associated with tool use in bottlenose dolphins (*Tursiops sp.*)? *Behavior Genetics* **40**, 706.

Baier, F. and Tokuriki, N. 2014. Connectivity between catalytic landscapes of the metallo-beta-lactamase superfamily. *Journal of Molecular Biology* **426**, 2442.

Balla, T., Szentpetery, Z. and Kim, Y.J. 2009. Phosphoinositide signaling: New tools and insights. *Physiology* **24**, 231.

Bansal, P. and Martin, A. 2000. Comparative study of vapour compression, thermoelectric and absorption refrigerators. *International Journal of Energy Research* **24**, 93.

Bar-On, Y.M., Phillips, R. and Milo, R. 2018. The biomass distribution on Earth. *Proceedings of the National Academy of Sciences of the United States of America* **115**, 6506.

Barton, N.H., Briggs, D.E.G., Eisen, J.A. et al. 2007. *Evolution.* Cold Spring Harbor Laboratory Press, Cold Spring Harbor, NY.

Barve, A. and Wagner, A. 2013. A latent capacity for evolutionary innovation through exaptation in metabolic systems. *Nature* **500**, 203.

Bedard, D.L., Haberl, M.L., May, R.J. et al. 1987. Evidence for novel mechanisms of polychlorinated biphenyl metabolism in *Alcaligenes eutrophus* H850. *Applied and Environmental Microbiology* **53**, 1103.

Bekpen, C., Marques-Bonet, T., Alkan, C. et al. 2009. Death and resurrection of the human IRGM gene. *PLOS Genetics* **5**, e1000403.

Bell, A. 28 February 2014. How the refrigerator got its hum. *The Guardian*.

Bell, E.A., Boehnke, P., Harrison, T.M. et al. 2015. Potentially biogenic carbon preserved in a 4.1 billion-year-old zircon. *Proceedings of the National Academy of Sciences of the United States of America* **112**, 14518.

Bell, M.A. and Aguirre, W.E. 2013. Contemporary evolution, allelic recycling, and adaptive radiation of the threespine stickleback. *Evolutionary Ecology Research* **15**, 377.

Bell, M.A., Aguirre, W.E. and Buck, N.J. 2004. Twelve years of contemporary armor evolution in a threespine stickleback population. *Evolution* **58**, 814.

Bell, M.A., Travis, M.P. and Blouw, D.M. 2006. Inferring natural selection in a fossil threespine stickleback. *Paleobiology* **32**, 562.

Benson-Amram, S., Gilfillan, G. and McComb, K. 2018. Numerical assessment in the wild: Insights from social carnivores. *Philosophical Transactions of the Royal Society B: Biological Sciences* **373**.

Berg, I.A., Kockelkorn, D., Buckel, W. et al. 2007. A 3-hydroxypropionate/4-hydroxybutyrate autotrophic carbon dioxide assimilation pathway in Archaea. *Science* **318**, 1782.

Bhanoo, S.N. 25 May 2015. Even Einstein's research can take time to matter. *The New York Times*.

Bhullar, K., Waglechner, N., Pawlowski, A. et al. 2012. Antibiotic resistance is prevalent in an isolated cave microbiome. *PLOS One* **7**.

Bird, C.D. and Emery, N.J. 2009. Rooks use stones to raise the water level to reach a floating worm. *Current Biology* **19**, 1410.

Blain, J.C. and Szostak, J.W. 2014. Progress toward synthetic cells. In *Annual Review of Biochemistry, Vol. 83* (ed. R.D. Kornberg), p. 615.

Blair, J.M.A., Webber, M.A., Baylay, A.J. et al. 2015. Molecular mechanisms of antibiotic resistance. *Nature Reviews Microbiology* **13**, 42.

Blount, Z.D. 2015. The natural history of model organisms: The unexhausted potential of *E. coli*. *eLife* **4**, e05826.

Blount, Z.D., Borland, C.Z. and Lenski, R.E. 2008. Historical contingency and the evolution of a key innovation in an experimental population of *Escherichia coli*. *Proceedings of the National Academy of Sciences of the United States of America* **105**, 7899.

Blount, Z.D., Lenski, R.E. and Losos, J.B. 2018. Contingency and determinism in evolution: Replaying life's tape. *Science* **362**, eaam5979.

Blüh, O. 1952. The value of inspiration. A study on Julius Robert Mayer and Josef Popper-Lynkeus. *Isis* **43**, 211.

Boesch, C., Marchesi, P., Marchesi, N. et al. 1994. Is nut cracking in wild chimpanzees a cultural behavior? *Journal of Human Evolution* **26**, 325.

Boraas, M.E., Seale, D.B. and Boxhorn, J.E. 1998. Phagotrophy by a flagellate selects for colonial prey: A possible origin of multicellularity. *Evolutionary Ecology* **12**, 153.

Bornman, L.J., Chris, H. and Botha, C. 1973. *Welwitschia mirabilis*: Observations on movement of water and assimilates under föhn and fog conditions. *Madoqua* **2**, 25.

Boroditsky, L. 2000. Metaphoric structuring: Understanding time through spatial metaphors. *Cognition* **75**, 1.

Bourke, A.F. 2014. Hamilton's rule and the causes of social evolution. *Philosophical Transactions of the Royal Society B: Biological Sciences* **369**, 20130362.

Bowdle, B.F. and Gentner, D. 2005. The career of metaphor. *Psychological Review* **112**, 193.

Bowern, C. and Zentz, J. 2012. Diversity in the numeral systems of Australian languages. *Anthropological Linguistics* **54**, 133.

Boyle, A.W., Silvin, C.J., Hassett, J.P. et al. 1992. Bacterial PCB biodegradation. *Biodegradation* **3**, 285.

Brawand, D., Wagner, C.E., Li, Y.I. et al. 2014. The genomic substrate for adaptive radiation in African cichlid fish. *Nature* **513**, 375.

Bremer, K. 2002. Gondwanan evolution of the grass alliance of families (Poales). *Evolution* **56**, 1374.

Brocklehurst, N., Panciroli, E., Benevento, G.L. et al. 2021. Mammaliaform extinctions as a driver of the morphological radiation of Cenozoic mammals. *Current Biology* **31**, 2955.

Brockman, J. 2017. *Know This. Today's Most Interesting and Important Scientific Ideas, Discoveries, and Developments*. Harper, New York, NY.

Brown, M.G. and Balkwill, D.L. 2009. Antibiotic resistance in bacteria isolated from the deep terrestrial subsurface. *Microbial Ecology* **57**, 484.

Bulliet, R.W. 2016. *The Wheel: Inventions and Reinventions*. Columbia University Press, New York, NY.

Burr, A.B. and Andrew, G.E. 1992. *The Unreasonable Effectiveness of Number Theory*. American Mathematical Society, Washington DC.

Bush, K. and Jacoby, G.A. 2010. Updated functional classification of β-lactamases. *Antimicrobial Agents and Chemotherapy* **54**, 969.

Bushman, F. 2002. *Lateral DNA Transfer: Mechanisms and Consequences*. Cold Spring Harbor University Press, Cold Spring Harbor, NY.

Butler, K. 18 June 2010. What broke my father's heart. *New York Times Magazine*.

Butterworth, B., Gallistel, C.R. and Vallortigara, G. 2018. Introduction: The origins of numerical abilities. *Philosophical Transactions of the Royal Society B: Biological Sciences* **373**, 20160507.

Bylund, E. and Athanasopoulos, P. 2017. The Whorfian time warp: Representing duration through the language hourglass. *Journal of Experimental Psychology: General* **146**, 911.

Cai, J., Zhao, R., Jiang, H. et al. 2008. *De novo* origination of a new protein-coding gene in Saccharomyces cerevisiae. *Genetics* **179**, 487.

Cairns, J., Overbaugh, J. and Miller, S. 1988. The origin of mutants. *Nature* **335**, 142.

Cameron, L. 2008. Metaphor and talk. In *The Cambridge Handbook of Metaphor and Thought* (ed. R.W. Gibbs Jr.), p. 197. Cambridge University Press, Cambridge, UK.

Campbell, D.T. 1960. Blind variation and selective retention in creative thought as in other knowledge processes. *Psychological Review* **67**, 380.

Cantlon, J.F. 2018. How evolution constrains human numerical concepts. *Child Development Perspectives* **12**, 65.

Cantrell, L. and Smith, L.B. 2013. Open questions and a proposal: A critical review of the evidence on infant numerical abilities. *Cognition* **128**, 331.

Caro, R.A. 1982. *The Path to Power: The Years of Lyndon Johnson I*. Knopf, New York, NY.

Carroll, L. 1871. *Through the Looking-Glass: And What Alice Found There*. Macmillan, United Kingdom.

Carvunis, A.-R., Rolland, T., Wapinski, I. et al. 2012. Proto-genes and *de novo* gene birth. *Nature* **487**, 370.

Casasanto, D. and Boroditsky, L. 2008. Time in the mind: Using space to think about time. *Cognition* **106**, 579.

Casola, C. 2018. From *de novo* to 'de nono': The majority of novel protein-coding genes identified with phylostratigraphy are old genes or recent duplicates. *Genome Biology and Evolution* **10**, 2906.

Catling, D.C., Glein, C.R., Zahnle, K.J. et al. 2005. Why O_2 is required by complex life on habitable planets and the concept of planetary 'oxygenation time'. *Astrobiology* **5**, 415.

Chang, K. 2 December 2008. A new picture of the early Earth. *The New York Times*.

Chang, Y.I. and Su, C.Y. 2003. Flocculation behavior of *Sphingobium chlorophenolicum* in degrading pentachlorophenol at different life stages. *Biotechnology and Bioengineering* **82**, 843.

Changizi, M.A. and Shimojo, S. 2005. Character complexity and redundancy in writing systems over human history. *Proceedings of the Royal Society B: Biological Sciences* **272**, 267.

Changizi, M.A., Zhang, Q., Ye, H. et al. 2006. The structures of letters and symbols throughout human history are selected to match those found in objects in natural scenes. *The American Naturalist* **167**, E117.

Chen, B.M., Liao, H.X., Chen, W.B. et al. 2017. Role of allelopathy in plant invasion and control of invasive plants. *Allelopathy Journal* **41**, 155.

Chen, S.D., Zhang, Y.E. and Long, M.Y. 2010. New genes in *Drosophila* quickly become essential. *Science* **330**, 1682.

Chernykh, E.N. 2008. Formation of the Eurasian 'steppe belt' of stockbreeding cultures: Viewed through the prism of archaeometallurgy and radiocarbon dating. *Archeology Ethnology & Anthropology of Eurasia* **35**, 36.

Christin, P.A., Besnard, G., Samaritani, E. et al. 2008. Oligocene CO_2 decline promoted C-4 photosynthesis in grasses. *Current Biology* **18**, 37.

Claramunt, S. and Cracraft, J. 2015. A new time tree reveals Earth history's imprint on the evolution of modern birds. *Science Advances* **1**, e1501005.

Clemente, J.C., Pehrsson, E.C., Blaser, M.J. et al. 2015. The microbiome of uncontacted Amerindians. *Science Advances* **1**, e1500183.

Colbourne, J.K., Pfrender, M.E., Gilbert, D. et al. 2011. The eco-responsive genome of *Daphnia pulex*. *Science* **331**, 555.

Cole, D.B., Mills, D.B., Erwin, D.H. et al. 2020. On the co-evolution of surface oxygen levels and animals. *Geobiology* **18**, 260.

Colosimo, P.F., Hosemann, K.E., Balabhadra, S. et al. 2005. Widespread parallel evolution in sticklebacks by repeated fixation of ectodysplasin alleles. *Science* **307**, 1928.

Constantinescu, A.O., O'Reilly, J.X. and Behrens, T.E. 2016. Organizing conceptual knowledge in humans with a gridlike code. *Science* **352**, 1464.

Conway Morris, S. 2003. *Life's Solution. Inevitable Humans in a Lonely Universe*. Cambridge University Press, New York, NY.

Cook, D.L., Gerber, L.N. and Tapscott, S.J. 1998. Modeling stochastic gene expression: Implications for haploinsufficiency. *Proceedings of the National Academy of Sciences of the United States of America* **95**, 15641.

Copley, S.D. 2017. Shining a light on enzyme promiscuity. *Current Opinion in Structural Biology* **47**, 167.

Copley, S.D., Rokicki, J., Turner, P. et al. 2012. The whole genome sequence of *Sphingobium chlorophenolicum L-1*: Insights into the evolution of the pentachlorophenol degradation pathway. *Genome Biology and Evolution* **4**, 184.

Coughenour, M.B. 1985. Graminoid responses to grazing by large herbivores – adaptations, exaptations, and interacting processes. *Annals of the Missouri Botanical Garden* **72**, 852.

Cowan, R.S. 1999. How the refrigerator got its hum. In *The Social Shaping of Technology* (eds. D. MacKenzie and J. Wajcman), p. 202. Open University, Buckingham.

Crofts, T.S., Gasparrini, A.J. and Dantas, G. 2017. Next-generation approaches to understand and combat the antibiotic resistome. *Nature Reviews Microbiology* 15, 422.

Csikszentmihalyi, M. 1996. *Creativity: The Psychology of Discovery and Invention.* HarperCollins, New York, NY.

D'Ari, R. and Casadesus, J. 1998. Underground metabolism. *Bioessays* 20, 181.

D'Costa, V.M., King, C.E., Kalan, L. et al. 2011. Antibiotic resistance is ancient. *Nature* 477, 457.

D'Errico, F., Doyon, L., Colage, I. et al. 2018. From number sense to number symbols. An archaeological perspective. *Philosophical Transactions of the Royal Society B: Biological Sciences* 373, 20160518.

Daane, J.M., Dornburg, A., Smits, P. et al. 2019. Historical contingency shapes adaptive radiation in Antarctic fishes. *Nature Ecology & Evolution* 3, 1102.

Dai, M.H. and Copley, S.D. 2004. Genome shuffling improves degradation of the anthropogenic pesticide pentachlorophenol by *Sphingobium chlorophenolicum* ATCC 39723. *Applied and Environmental Microbiology* 70, 2391.

Dannen, G. 1997. The Einstein-Szilard refrigerators. *Scientific American* 276, 90.

Dantas, G., Sommer, M.O.A., Oluwasegun, R.D. et al. 2008. Bacteria subsisting on antibiotics. *Science* 320, 100.

Darwin, C. 1872. *The Origin of Species by Means of Natural Selection; or The Preservation of Favored Races in the Struggle for Life* (6th ed., reprinted by A.L. Burt, New York). John Murray London, England.

Davenport, D. 2012. The war against bacteria: How were sulphonamide drugs used by Britain during World War II? *Medical Humanities* 38, 55.

Davies, J. and Davies, D. 2010. Origins and evolution of antibiotic resistance. *Microbiology and Molecular Biology Reviews* 74, 417.

Deamer, D.W. 1998. Membrane compartments in prebiotic evolution. In *The Molecular Origins of Life: Assembling Pieces of the Puzzle* (ed. A. Brack), p. 189. Cambridge University Press, Cambridge, UK.

Dean, B.P. and Scharnhorst, G. 1990. The contemporary reception of 'Walden'. In *Studies in the American Renaissance 1990* (ed. J. Myerson), p. 293. University of Virginia Press, Charlottesville.

Dehaene, S. 2009. *Reading in the Brain. The New Science of How We Read.* Penguin, New York, NY.

Dehaene, S. 2011. *The Number Sense. How The Mind Creates Mathematics.* Oxford University Press, New York, NY.

Dehaene, S. and Changeux, J.P. 1993. Development of elementary numerical abilities – a neuronal model. *Journal of Cognitive Neuroscience* 5, 390.

Dehaene, S. and Cohen, L. 2007. Cultural recycling of cortical maps. *Neuron* 56, 384.

Dehaene, S. and Cohen, L. 2011. The unique role of the visual word form area in reading. *Trends in Cognitive Sciences* 15, 254.

Dehaene, S., Pegado, F., Braga, L.W. et al. 2010. How learning to read changes the cortical networks for vision and language. *Science* 330, 1359.

Denef, V.J., Park, J., Tsoi, T.V. et al. 2004. Biphenyl and benzoate metabolism in a genomic context: Outlining genome-wide metabolic networks in *Burkholderia xenovorans* LB400. *Applied and Environmental Microbiology* 70, 4961.

Denoeud, F., Carretero-Paulet, L., Dereeper, A. et al. 2014. The coffee genome provides insight into the convergent evolution of caffeine biosynthesis. *Science* 345, 1181.

Desmond, E. and Gribaldo, S. 2009. Phylogenomics of sterol synthesis: Insights into the origin, evolution, and diversity of a key eukaryotic feature. *Genome Biology and Evolution* 1, 364.

Dhar, R., Sägesser, R., Weikert, C. et al. 2013. Yeast adapts to a changing stressful environment by evolving cross-protection and anticipatory gene regulation. *Molecular Biology and Evolution* **30**, 573.

Dhar, R., Sagesser, R., Weikert, C. et al. 2011. Adaptation of *Saccharomyces cerevisiae* to saline stress through laboratory evolution. *Journal of Evolutionary Biology* **24**, 1135.

Diester, I. and Nieder, A. 2007. Semantic associations between signs and numerical categories in the prefrontal cortex. *PLOS Biology* **5**, 2684.

Dillon, G. and Millay, E.S.V. 1936. *Flowers of Evil; from the French of Charles Baudelaire*. Harper & Brothers, New York, NY.

Dinan, L. 2001. Phytoecdysteroids: Biological aspects. *Phytochemistry* **57**, 325.

Diogo, R., Richmond, B.G. and Wood, B. 2012. Evolution and homologies of primate and modern human hand and forearm muscles, with notes on thumb movements and tool use. *Journal of Human Evolution* **63**, 64.

Dodd, M.S., Papineau, D., Grenne, T. et al. 2017. Evidence for early life in Earth's oldest hydrothermal vent precipitates. *Nature* **543**, 60.

Domazet-Loso, T., Carvunis, A.R., Alba, M.M. et al. 2017. No evidence for phylostratigraphic bias impacting inferences on patterns of gene emergence and evolution. *Molecular Biology and Evolution* **34**, 843.

Du, D.J., Wang-Kan, X., Neuberger, A. et al. 2018. Multidrug efflux pumps: Structure, function and regulation. *Nature Reviews Microbiology* **16**, 523.

Du, J. and Wu, Y.S. 2018. A parameter-free index for identifying under-cited sleeping beauties in science. *Scientometrics* **116**, 959.

Duarte, N.C., Becker, S.A., Jamshidi, N. et al. 2007. Global reconstruction of the human metabolic network based on genomic and bibliomic data. *Proceedings of the National Academy of Sciences* **104**, 1777.

Edwards, E.J. and Smith, S.A. 2010. Phylogenetic analyses reveal the shady history of C-4 grasses. *Proceedings of the National Academy of Sciences of the United States of America* **107**, 2532.

Eger, E., Sterzer, P., Russ, M.O. et al. 2003. A supramodal number representation in human intraparietal cortex. *Neuron* **37**, 719.

El Aichouchi, A. and Gorry, P. 2018. Delayed recognition of Judah Folkman's hypothesis on tumor angiogenesis: When a Prince awakens a Sleeping Beauty by self-citation. *Scientometrics* **116**, 385.

ENCODE Project Consortium. 2012. An integrated encyclopedia of DNA elements in the human genome. *Nature* **489**, 57.

Epps, P., Bowern, C., Hansen, C.A. et al. 2012. On numeral complexity in hunter-gatherer languages. *Linguistic Typology* **16**, 41.

Erickson, G.M., Rauhut, O.W., Zhou, Z. et al. 2009. Was dinosaurian physiology inherited by birds? Reconciling slow growth in Archaeopteryx. *PLOS One* **4**, e7390.

Erwin, D.H. 1992. A preliminary classification of evolutionary radiations. *Historical Biology* **6**, 133.

Erwin, D.H., Laflamme, M., Tweedt, S.M. et al. 2011. The Cambrian conundrum: Early divergence and later ecological success in the early history of animals. *Science* **334**, 1091.

Erwin, D.H. and Valentine, J.W. 2013. *The Cambrian Explosion. The Construction of Animal Biodiversity*. Roberts and Company, Greenwood Village, CO.

Estes, S. and Arnold, S.J. 2007. Resolving the paradox of stasis: Models with stabilizing selection explain evolutionary divergence on all timescales. *The American Naturalist* **169**, 227.

Falótico, T., Proffitt, T., Ottoni, E.B. et al. 2019. Three thousand years of wild capuchin stone tool use. *Nature Ecology & Evolution* **3**, 1034.

Farrell, B.D., Dussourd, D.E. and Mitter, C. 1991. Escalation of plant defense – do latex and resin canals spur plant diversification? *The American Naturalist* **138**, 881.

Feduccia, A. 2003. 'Big bang' for tertiary birds? *Trends in Ecology & Evolution* **18**, 172.

Feigenson, L., Carey, S. and Spelke, E. 2002. Infants' discrimination of number vs. continuous extent. *Cognitive Psychology* **44**, 33.

Feigenson, L., Libertus, M.E. and Halberda, J. 2013. Links between the intuitive sense of number and formal mathematics ability. *Child Development Perspectives* **7**, 74.

Feist, A.M., Henry, C.S., Reed, J.L. et al. 2007. A genome-scale metabolic reconstruction for *Escherichia coli* K-12 MG1655 that accounts for 1260 ORFs and thermodynamic information. *Molecular Systems Biology* **3**.

Felsenstein, J. 2004. *Inferring Phylogenies.* Sinauer Associates, Sunderland, Massachusetts.

Fenchel, T. and Finlay, B.J. 1995. *Ecology and Evolution in Anoxic Worlds.* Oxford University Press, Oxford, UK.

Ferrigno, S., Jara-Ettinger, J., Piantadosi, S.T. et al. 2017. Universal and uniquely human factors in spontaneous number perception. *Nature Communications* **8**.

Field, D.J., Bercovici, A., Berv, J.S. et al. 2018. Early evolution of modern birds structured by global forest collapse at the end-Cretaceous mass extinction. *Current Biology* **28**, 1825.

Finch, B. 2015. The true story of Kudzu, the vine that never truly ate the South. *Smithsonian Magazine.*

Flexner, A. 1939 The usefulness of useless knowledge. *Harpers* **179**, 544.

Fox, D. 2016. What sparked the Cambrian explosion? *Nature* **530**, 268.

Fox, E.A., van Schaik, C.P., Sitompul, A. et al. 2004. Intra- and interpopulational differences in orangutan (*Pongo pygmaeus*) activity and diet: Implications for the invention of tool use. *American Journal of Physical Anthropology* **125**, 162.

Frank, D.A., McNaughton, S.J. and Tracy, B.F. 1998. The ecology of the Earth's grazing ecosystems. *Bioscience* **48**, 513.

Frazier, I. 2005. Destroying Baghdad. *New Yorker* 25.

Fujii, J.A., Ralls, K. and Tinker, M.T. 2015. Ecological drivers of variation in tool-use frequency across sea otter populations. *Behavioral Ecology* 26, 519.

Futuyma, D.J. 1998. *Evolutionary Biology*. Sinauer, Sunderland, Massachusetts.

Gallagher, P. 2 July 2015. Forget little green men – aliens will look like humans, says Cambridge University evolution expert. *The Independent*.

Galton, D. 2009. Did Darwin read Mendel? *QJM: An International Journal of Medicine* 102, 587.

Gamow, G. 1966. *Thirty Years That Shook Physics. The Story of Quantum Theory*. Doubleday, Garden City, NY.

Gancedo, C. and Flores, C.L. 2008. Moonlighting proteins in yeasts. *Microbiology and Molecular Biology Reviews* 72, 197.

Gärdenfors, P. 2000. *Conceptual Spaces. The Geometry of Thought*. MIT Press, Cambridge, MA.

Garfield, E. 1989a. Delayed recognition in scientific discovery – citation frequency analysis and the search for case histories. *Current Contents* 23, 3.

Garfield, E. 1989b. More delayed recognition. Part 1. Examples from the genetics of color-blindness, the entropy of short-term-memory, phosphoinositides, and polymer rheology. *Current Contents* 38, 3.

Garfield, E. 1990. More delayed recognition. Part 2. From inhibin to scanning electron microscopy. *Current Contents* 9, 3.

Geist, D.J., Snell, H., Snell, H. et al. 2014. A paleogeographic model of the Galápagos Islands and biogeographical and evolutionary implications. In *The Galápagos: A Natural Laboratory for the Earth Sciences* (eds. K.S. Harpp, E. Mittelstaedt, N. d'Ozouville and D.W. Graham), p. 145. American Geophysical Union, Washington DC.

Genner, M.J., Seehausen, O., Lunt, D.H. et al. 2007. Age of cichlids: New dates for ancient lake fish radiations. *Molecular Biology and Evolution* **24**, 1269.

Gentner, D. 2010. Bootstrapping the mind: Analogical processes and symbol systems. *Cognitive Science* **34**, 752.

Gentner, D. and Hoyos, C. 2017. Analogy and abstraction. *Topics in Cognitive Science* **9**, 672.

Gentner, D. and Jeziorski, M. 1989. Historical shifts in the use of analogy in science. In *Psychology of Science: Contributions to Metascience* (eds. B.E. Gholson, W.R. Shadish Jr, R.A. Neimeyer and A.C. Houts). Cambridge University Press, New York, NY.

Gerhart, J. and Kirschner, M. 1998. *Cells, Embryos, and Evolution. Toward a Cellular and Developmental Understanding of Phenotypic Variation and Evolutionary Adaptability*. Blackwell, Boston, MA.

Gibbs, M.A. and Hosea, N.A. 2003. Factors affecting the clinical development of cytochrome P450 3A substrates. *Clinical Pharmacokinetics* **42**, 969.

Gibson, G. 2005. The synthesis and evolution of a supermodel. *Science* **307**, 1890.

Giles, N. 1983. The possible role of environmental calcium levels during the evolution of phenotypic diversity in Outer Hebridean populations of the three-spined stickleback, *Gasterosteus aculeatus. Journal of Zoology* **199**, 535.

Gould, S. and Vrba, E. 1982. Exaptation – a missing term in the science of form. *Paleobiology* **8**, 4.

Gould, S.J. 1990. *Wonderful Life: The Burgess Shale and the Nature of History*. W.W. Norton & Company, New York, NY.

Gould, S.J. and Eldredge, N. 1977. Punctuated equilibria: Tempo and mode of evolution reconsidered. *Paleobiology* **3**, 115.

Gramling, C. 2014. Low oxygen stifled animals' emergence, study says. *Science* **346**, 537.

Graur, D., Zheng, Y., Price, N. et al. 2013. On the immortality of television sets: 'Function' in the human genome according to the evolution-free gospel of ENCODE. *Genome Biology and Evolution* **5**, 578.

Graves Jr, J., Hertweck, K., Phillips, M. et al. 2017. Genomics of parallel experimental evolution in *Drosophila*. *Molecular Biology and Evolution* **34**, 831.

Grebenok, R.J., Galbraith, D.W., Benveniste, I. et al. 1996. Ecdysone 20-monooxygenase, a cytochrome P450 enzyme from spinach, *Spinacia oleracea*. *Phytochemistry* **42**, 927.

Grimaldi, D. and Agosti, D. 2000. A formicine in New Jersey Cretaceous amber (Hymenoptera: Formicidae) and early evolution of the ants. *Proceedings of the National Academy of Sciences of the United States of America* **97**, 13678.

Grossnickle, D.M., Smith, S.M. and Wilson, G.P. 2019. Untangling the multiple ecological radiations of early mammals. *Trends in Ecology & Evolution* **34**, 936.

Grunwald, M. 1 January 2002. Monsanto hid decades of pollution. *The Washington Post*.

Guterl, F. 1994. Suddenly, number theory makes sense to industry. *Math Horizons* **2**, 6.

Halberda, J., Mazzocco, M.M.M. and Feigenson, L. 2008. Individual differences in non-verbal number acuity correlate with maths achievement. *Nature* **455**, 665.

Hall, B.G. 1982. Chromosomal mutation for citrate utilization by *Escherichia coli* K-12. *Journal of Bacteriology* **151**, 269.

Hall, D.A., Zhu, H., Zhu, X.W. et al. 2004. Regulation of gene expression by a metabolic enzyme. *Science* **306**, 482.

Halligan, D.L. and Keightley, P.D. 2006. Ubiquitous selective constraints in the *Drosophila* genome revealed by a genome-wide interspecies comparison. *Genome Research* **16**, 875.

Hansen, T.F. and Houle, D. 2004. Evolvability, stabilizing selection, and the problem of stasis. In *Phenotypic Integration:*

Studying the Ecology and Evolution of Complex Phenotypes (eds. M. Pigliucci and K. Preston), p. 130. Oxford University Press, Oxford, UK.

Hardy, G.H. 1967. *A Mathematician's Apology*. Cambridge University Press, Cambridge, UK.

Hartl, D.L. and Clark, A.G. 2007. *Principles of Population Genetics*. Sinauer Associates, Sunderland, MA.

Hartleb, D., Jarre, F. and Lercher, M.J. 2016. Improved metabolic models for *E. coli* and *Mycoplasma genitalium* from GlobalFit, an algorithm that simultaneously matches growth and non-growth data sets. *PLOS Computational Biology* **12**, e1005036.

Haslam, M. 2013. 'Captivity bias' in animal tool use and its implications for the evolution of hominin technology. *Philosophical Transactions of the Royal Society B: Biological Sciences* **368**.

Haslam, M., Luncz, L.V., Staff, R.A. et al. 2016. Pre-Columbian monkey tools. *Current Biology* **26**, R521.

Hauser, M.D., Chomsky, N. and Fitch, W.T. 2002. The faculty of language: What is it, who has it, and how did it evolve? *Science* **298**, 1569.

Hausman, C. and Kellogg, R. 2015. Welfare and distributional implications of shale gas (Working Paper No. 21115). National Bureau of Economic Research.

Heard, S.B. and Hauser, D.L. 1995. Key evolutionary innovations and their ecological mechanisms. *Historical Biology* **10**, 151.

Hendrickson, H., Slechta, E.S., Bergthorsson, U. et al. 2002. Amplification–mutagenesis: Evidence that "directed" adaptive mutation and general hypermutability result from growth with a selected gene amplification. *Proceedings of the National Academy of Sciences* **99**, 2164.

Hiratsuka, T., Furihata, K., Ishikawa, J. et al. 2008. An alternative menaquinone biosynthetic pathway operating in microorganisms. *Science* **321**, 1670.

Hlouchova, K., Rudolph, J., Pietari, J.M.H. et al. 2012. Pentachlorophenol hydroxylase, a poorly functioning enzyme required for degradation of pentachlorophenol by *Sphingobium chlorophenolicum*. *Biochemistry* **51**, 3848.

Ho, S.Y. and Duchêne, S. 2014. Molecular-clock methods for estimating evolutionary rates and timescales. *Molecular Ecology* **23**, 5947.

Hokin, M.R. and Hokin, L.E. 1953. Enzyme secretion and the incorporation of P-32 into phospholipides of pancreas slices. *Journal of Biological Chemistry* **203**, 967.

Hosler, D., Burkett, S.L. and Tarkanian, M.J. 1999. Prehistoric polymers: Rubber processing in ancient Mesoamerica. *Science* **284**, 1988.

Howard, S.R., Avarguès-Weber, A., Garcia, J.E. et al. 2018. Numerical ordering of zero in honey bees. *Science* **360**, 1124.

Hoyle, F. 1950. *The Nature of the Universe*. Basil Blackwell, Oxford, UK.

Huang, H., Pandya, C., Liu, C. et al. 2015. Panoramic view of a superfamily of phosphatases through substrate profiling. *Proceedings of the National Academy of Sciences of the United States of America* **112**, E1974.

Huang, R.Q., O'Donnell, A.J., Barboline, J.J. et al. 2016. Convergent evolution of caffeine in plants by co-option of exapted ancestral enzymes. *Proceedings of the National Academy of Sciences of the United States of America* **113**, 10613.

Huffman, M.A. 2003. Animal self-medication and ethno-medicine: Exploration and exploitation of the medicinal properties of plants. *Proceedings of the Nutrition Society* **62**, 371.

Huffman, M.A., Nahallage, C.A.D. and Leca, J.B. 2008. Cultured monkeys: Social learning cast in stones. *Current Directions in Psychological Science* **17**, 410.

Hughes, C. and Eastwood, R. 2006. Island radiation on a continental scale: Exceptional rates of plant diversification after uplift of the Andes. *Proceedings of the National Academy of Sciences of the United States of America* **103**, 10334.

Hume, D. 1740. *An Abstract of a Book lately Published; entituled, A Treatise of Human Nature &c. Wherein the Chief Argument of that Book is farther Illustrated and Explained.* C. Borbet, London.

Hunt, G. 2007. The relative importance of directional change, random walks, and stasis in the evolution of fossil lineages. *Proceedings of the National Academy of Sciences of the United States of America* **104**, 18404.

Hunt, G., Bell, M.A. and Travis, M.P. 2008. Evolution toward a new adaptive optimum: Phenotypic evolution in a fossil stickleback lineage. *Evolution* **62**, 700.

Hunt, G., Hopkins, M.J. and Lidgard, S. 2015. Simple versus complex models of trait evolution and stasis as a response to environmental change. *Proceedings of the National Academy of Sciences* **112**, 4885.

Hunt, G. and Rabosky, D.L. 2014. Phenotypic evolution in fossil species: Pattern and process. *Annual Review of Earth and Planetary Sciences* **42**, 421.

Hunt, G.R. and Gray, R.D. 2003. Diversification and cumulative evolution in New Caledonian crow tool manufacture. *Proceedings of the Royal Society B: Biological Sciences* **270**, 867.

Hunt, G.R. and Gray, R.D. 2004. The crafting of hook tools by wild New Caledonian crows. *Proceedings of the Royal Society B: Biological Sciences* **271**, S88.

Hunter, J.P. and Jernvall, J. 1995. The hypocone as a key innovation in mammalian evolution. *Proceedings of the National Academy of Sciences of the United States of America* **92**, 10718.

Inderjit. 2012. Exotic plant invasion in the context of plant defense against herbivores. *Plant Physiology* **158**, 1107.

Iyer, M.K., Niknafs, Y.S., Malik, R. et al. 2015. The landscape of long noncoding RNAs in the human transcriptome. *Nature Genetics* **47**, 199.

Jablonski, D. 2017. Approaches to macroevolution: 1. General concepts and origin of variation. *Evolutionary Biology* **44**, 427.

Jacob, F. 1977. Evolution and tinkering. *Science* **196**, 1161.

Ji, Q., Luo, Z.X., Yuan, C.X. et al. 2006. A swimming mammaliaform from the Middle Jurassic and ecomorphological diversification of early mammals. *Science* **311**, 1123.

Ji, Q., Luo, Z.X., Yuan, C.X. et al. 2002. The earliest known eutherian mammal. *Nature* **416**, 816.

Judson, O.P. 2017. The energy expansions of evolution. *Nature Ecology & Evolution* **1**, 1–9.

Kandel, E.R., Schwartz, J.H., Jessell, T.M. et al. 2013. *Principles of Neural Science*. McGraw Hill, New York, NY.

Kanehisa, M., Furumichi, M., Tanabe, M. et al. 2016. KEGG: New perspectives on genomes, pathways, diseases and drugs. *Nucleic Acids Research* **45**, D353.

Kardos, N. and Demain, A.L. 2011. Penicillin: The medicine with the greatest impact on therapeutic outcomes. *Applied Microbiology and Biotechnology* **92**, 677.

Karve, S. and Wagner, A. 2022a. Multiple novel traits without immediate benefits originate in bacteria evolving on single antibiotics. *Molecular Biology and Evolution* **39**, msab341.

Karve, S. and Wagner, A. 2022b. Environmental complexity is more important than mutation in driving the evolution of latent novel traits in *E. coli*. *Nature Communications* (in press).

Kawai, J. 1965. Newly-acquired pre-cultural behavior of the natural troop of Japanese monkeys on Koshima islet. *Primates* **6**, 1.

Ke, Q., Ferrara, E., Radicchi, F. et al. 2015. Defining and identifying Sleeping Beauties in science. *Proceedings of the National Academy of Sciences of the United States of America* **112**, 7426.

Kellis, M., Birren, B.W. and Lander, E.S. 2004. Proof and evolutionary analysis of ancient genome duplication in the yeast *Saccharomyces cerevisiae*. *Nature* **428**, 617.

Kellis, M., Patterson, N., Endrizzi, M. et al. 2003. Sequencing and comparison of yeast species to identify genes and regulatory elements. *Nature* **423**, 241.

Kenward, B., Schloegl, C., Rutz, C. et al. 2011. On the evolutionary and ontogenetic origins of tool-oriented behaviour in New Caledonian crows (*Corvus moneduloides*). *Biological Journal of the Linnean Society* **102**, 870.

Khalturin, K., Hemmrich, G., Fraune, S. et al. 2009. More than just orphans: Are taxonomically-restricted genes important in evolution? *Trends in Genetics* **25**, 404.

Khersonsky, O. and Tawfik, D.S. 2010. Enzyme promiscuity: A mechanistic and evolutionary perspective. *Annual Review of Biochemistry* **79**, 471.

Kilgour, F.G. 1963. Vitruvius and the early history of wave theory. *Technology and Culture* **4**, 282.

Kim, K.M., Qin, T., Jiang, Y.-Y. et al. 2012. Protein domain structure uncovers the origin of aerobic metabolism and the rise of planetary oxygen. *Structure* **20**, 67.

Kimura, F., Sato, M. and Kato-Noguchi, H. 2015. Allelopathy of pine litter: Delivery of allelopathic substances into forest floor. *Journal of Plant Biology* **58**, 61.

Knoll, A.H. 2011. The multiple origins of complex multicellularity. In *Annual Review of Earth and Planetary Sciences, Vol. 39* (eds. R. Jeanloz and K.H. Freeman), p. 217.

Knopp, M., Gudmundsdottir, J.S., Nilsson, T. et al. 2019. *De novo* emergence of peptides that confer antibiotic resistance. *MBio* **10**, e00837.

Kocher, T.D. 2004. Adaptive evolution and explosive speciation: The cichlid fish model. *Nature Reviews Genetics* **5**, 288.

Koehn, E.M., Fleischmann, T., Conrad, J.A. et al. 2009. An unusual mechanism of thymidylate biosynthesis in organisms containing the thyX gene. *Nature* **458**, 919.

Koestler, A. 1964. *The Act of Creation. A study of the conscious and unconscious processes of humor, scientific discovery and art.* Macmillan, New York, NY.

Koops, K., McGrew, W.C. and Matsuzawa, T. 2013. Ecology of culture: Do environmental factors influence foraging tool use in wild chimpanzees, *Pan troglodytes verus*? *Animal Behaviour* **85**, 175.

Koops, K., Visalberghi, E. and van Schaik, C.P. 2014. The ecology of primate material culture. *Biology Letters* **10**, 20140508.

Koschwanez, J.H., Foster, K.R. and Murray, A.W. 2013. Improved use of a public good selects for the evolution of undifferentiated multicellularity. *eLife* **2**, e00367.

Kottler, M.J. 1979. Hugo de Vries and the rediscovery of Mendel's laws. *Annals of Science* **36**, 517.

Kröger, B. and Penny, A. 2020. Skeletal marine animal biodiversity is built by families with long macroevolutionary lag times. *Nature Ecology & Evolution* **4**, 1410.

Krützen, M., Mann, J., Heithaus, M.R. et al. 2005. Cultural transmission of tool use in bottlenose dolphins. *Proceedings of the National Academy of Sciences of the United States of America* **102**, 8939.

Kutter, C., Watt, S., Stefflova, K. et al. 2012. Rapid turnover of long noncoding RNAs and the evolution of gene expression. *PLOS Genetics* **8**.

Kutter, E.F., Bostroem, J., Elger, C.E. et al. 2018. Single neurons in the human brain encode numbers. *Neuron* **100**, 753.

Ladoukakis, E., Pereira, V., Magny, E.G. et al. 2011. Hundreds of putatively functional small open reading frames in *Drosophila*. *Genome Biology* **12**, R118.

Lakoff, G. 1993. The contemporary theory of metaphor. In *Metaphor and Thought* (ed. A. Ortony). Cambridge University Press, New York, NY.

Lakoff, G. and Johnson, M. 1980. *Metaphors We Live By*. University of Chicago Press, Chicago, IL.

Lamb, J. 2001. *Preserving the Self in the South Seas, 1680–1840.* The University of Chicago Press, Chicago, IL.

Lane, N. and Martin, W. 2010. The energetics of genome complexity. *Nature* **467**, 929.

Lang, G.I., Rice, D.P., Hickman, M.J. et al. 2013. Pervasive genetic hitchhiking and clonal interference in forty evolving yeast populations. *Nature* **500**, 571.

Larson, G., Piperno, D.R., Allaby, R.G. et al. 2014. Current perspectives and the future of domestication studies. *Proceedings of the National Academy of Sciences of the United States of America* **111**, 6139.

Larsson, H.C.E., Hone, D.W., Dececchi, T.A. et al. 2010. The winged non-avian dinosaur *Microraptor* fed on mammals: Implications for the Jehol Biota ecosystem. *Journal of Vertebrate Paleontology* **30**, 114A.

Law, R. 1980. Wheeled transport in pre-colonial West Africa. *Africa* **50**, 249.

Leigh, M.B., Prouzova, P., Mackova, M. et al. 2006. Polychlorinated biphenyl (PCB)-degrading bacteria associated with trees in a PCB-contaminated site. *Applied and Environmental Microbiology* **72**, 2331.

Lenski, R.E. 2017. Experimental evolution and the dynamics of adaptation and genome evolution in microbial populations. *ISME Journal* **11**, 2181.

Lenski, R.E. and Mittler, J.E. 1993. The directed mutation controversy and neo-Darwinism. *Science* **259**, 188.

Leon, D., D'Alton, S., Quandt, E.M. et al. 2018. Innovation in an *E. coli* evolution experiment is contingent on maintaining adaptive potential until competition subsides. *PLOS Genetics* **14**, e1007348.

Levine, M.T., Jones, C.D., Kern, A.D. et al. 2006. Novel genes derived from noncoding DNA in *Drosophila melanogaster* are frequently X-linked and exhibit testis-biased expression. *Proceedings of the National Academy of Sciences of the United States of America* **103**, 9935.

Lewis, M.J.T. 1994. The origins of the wheelbarrow. *Technology and Culture* **35**, 453.

Li, J. and Shi, D. 2016. Sleeping beauties in genius work: When were they awakened? *Journal of the Association for Information Science and Technology* **67**, 432.

Libby, E., Hébert-Dufresne, L., Hosseini, S.-R. et al. 2019. Syntrophy emerges spontaneously in complex metabolic systems. *PLOS Computational Biology* **15**, e1007169.

Liem, K.F. 1973. Evolutionary strategies and morphological innovations – cichlid pharyngeal jaws. *Systematic Zoology* **22**, 425.

Lohr, S. 20 March 2007. John W. Backus, 82, Fortran developer, dies. *The New York Times*.

Losos, J.B. 2017. *Improbable Destinies. How Predictable is Evolution?* Riverhead Books, New York, NY.

Losos, J.B. and Mahler, D.L. 2010. Adaptive radiation: The interaction of ecological opportunity, adaptation, and speciation. In *Evolution Since Darwin: The First 150 Years* (eds. M.A. Bell, D.J. Futuyma, W.F. Eanes and J.S. Levinton), p. 381. Sinauer Associates, Sunderland, MA.

Luo, Z.-X. and Wible, J.R. 2005. A Late Jurassic digging mammal and early mammalian diversification. *Science* **308**, 103.

Luo, Z.-X., Yuan, C.-X., Meng, Q.-J. et al. 2011. A Jurassic eutherian mammal and divergence of marsupials and placentals. *Nature* **476**, 442.

Luo, Z.X. 2007. Transformation and diversification in early mammal evolution. *Nature* **450**, 1011.

Lynch, M. 2007. *The Origins of Genome Architecture.* Sinauer, Sunderland, MA.

Lynch, M. and Conery, J.S. 2000. The evolutionary fate and consequences of duplicate genes. *Science* **290**, 1151.

Lynch, M. and Marinov, G.K. 2015. The bioenergetic costs of a gene. *Proceedings of the National Academy of Sciences of the United States of America* **112**, 15690.

Lynch, M. and Walsh, B. 1998. *Genetics and Analysis of Quantitative Traits.* Sinauer, Sunderland, MA.

Lyons, N.A. and Kolter, R. 2015. On the evolution of bacterial multicellularity. *Current Opinion in Microbiology* **24**, 21.

Maan, M.E. and Sefc, K.M. 2013. Colour variation in cichlid fish: Developmental mechanisms, selective pressures and evolutionary consequences. *Seminars in Cell & Developmental Biology* **24**, 516.

MacFadden, B.J. 2005. Fossil horses – Evidence for evolution. *Science* **307**, 1728.

Mann, J. and Patterson, E.M. 2013. Tool use by aquatic animals. *Philosophical Transactions of the Royal Society B: Biological Sciences* **368**, 20120424.

Mann, J., Sargeant, B.L., Watson-Capps, J.J. et al. 2008. Why do Dolphins carry sponges? *PLOS One* **3**.

Marche, S. 25 July 2015. Failure is our muse. *The New York Times.*

Martin, S. and Drijfhout, F. 2009. A review of ant cuticular hydrocarbons. *Journal of Chemical Ecology* **35**, 1151.

Martins, J., Teles, L.O. and Vasconcelos, V. 2007. Assays with *Daphnia magna* and *Danio rerio* as alert systems in aquatic toxicology. *Environment International* **33**, 414.

Mazzocco, M.M.M., Feigenson, L. and Halberda, J. 2011. Impaired acuity of the approximate number system underlies mathematical learning disability (dyscalculia). *Child Development* **82**, 1224.

McGee, M.D., Borstein, S.R., Neches, R.Y. et al. 2015. A pharyngeal jaw evolutionary innovation facilitated extinction in Lake Victoria cichlids. *Science* **350**, 1077.

McGrew, W.C., Ham, R.M., White, L.J.T. et al. 1997. Why don't chimpanzees in Gabon crack nuts? *International Journal of Primatology* **18**, 353.

McKay, C. 1999. The Bach reception in the 18th and 19th centuries. http://www.music.mcgill.ca/~cmckay/

McKinnon, J.S. and Rundle, H.D. 2002. Speciation in nature: The threespine stickleback model systems. *Trends in Ecology & Evolution* **17**, 480.

McLysaght, A. and Guerzoni, D. 2015. New genes from non-coding sequence: The role of *de novo* protein-coding genes in eukaryotic evolutionary innovation. *Philosophical Transactions of the Royal Society B: Biological Sciences* **370**, 20140332.

Meader, S., Ponting, C.P. and Lunter, G. 2010. Massive turnover of functional sequence in human and other mammalian genomes. *Genome Research* **20**, 1335.

Meier, J.I., Marques, D.A., Mwaiko, S. et al. 2017. Ancient hybridization fuels rapid cichlid fish adaptive radiations. *Nature Communications* **8**, 1–11.

Meijnen, J.P., de Winde, J.H. and Ruijssenaars, H.J. 2008. Engineering *Pseudomonas putida* S12 for efficient utilization of D-xylose and L-arabinose. *Applied and Environmental Microbiology* **74**, 5031.

Mendel, G. 1866. Versuche über Pflanzen-Hybriden. *Verhandlungen des Naturforschenden Vereins Brünn* **4**, 3.

Meng, J., Hu, Y.M., Wang, Y.Q. et al. 2006. A mesozoic gliding mammal from northeastern China. *Nature* **444**, 889.

Mercader, J., Barton, H., Gillespie, J. et al. 2007. 4,300-year-old chimpanzee sites and the origins of percussive stone technology. *Proceedings of the National Academy of Sciences* **104**, 3043.

Merton, R.K. 1961. Singletons and multiples in scientific discovery: A chapter in the sociology of science. *Proceedings of the American Philosophical Society* **105**, 470.

Miller, S. 1953. A production of amino acids under possible primitive Earth conditions. *Science* **117**, 528.

Miller, S.L. 1998. The endogenous synthesis of organic compounds. In *The Molecular Origins of Life: Assembling Pieces of the Puzzle* (ed. A. Brack), p. 59. Cambridge University Press, Cambridge, UK.

Mills, D.B. and Canfield, D.E. 2014. Oxygen and animal evolution: Did a rise of atmospheric oxygen "trigger" the origin of animals? *Bioessays* **36**, 1145.

Mithofer, A. and Boland, W. 2012. Plant defense against herbivores: Chemical aspects. *Annual Review of Plant Biology* **63**, 431.

Mitter, C., Farrell, B. and Wiegmann, B. 1988. The phylogenetic study of adaptive zones – has phytophagy promoted insect diversification? *The American Naturalist* **132**, 107.

Mizutani, M. and Sato, F. 2011. Unusual P450 reactions in plant secondary metabolism. *Archives of Biochemistry and Biophysics* **507**, 194.

Mooney, C. 2011. The truth about fracking. *Scientific American* **305**, 80.

Moreau, C.S., Bell, C.D., Vila, R. et al. 2006. Phylogeny of the ants: Diversification in the age of angiosperms. *Science* **312**, 101.

Morris, B.E., Henneberger, R., Huber, H. et al. 2013. Microbial syntrophy: Interaction for the common good. *FEMS Microbiology Reviews* **37**, 384.

Morton, M.Q. 2013. Unlocking the Earth: A short history of hydraulic fracturing. *GEO Expro* **10**.

Moser, E.I., Moser, M.-B. and Roudi, Y. 2014. Network mechanisms of grid cells. *Philosophical Transactions of the Royal Society B: Biological Sciences* **369**, 20120511.

Mosteller, F. 1981. Innovation and evaluation. *Science* **211**, 881.

Moyers, B.A. and Zhang, J.Z. 2018. Toward reducing phylostratigraphic errors and biases. *Genome Biology and Evolution* **10**, 2037.

Muller, T.A., Werlen, C., Spain, J. et al. 2003. Evolution of a chlorobenzene degradative pathway among bacteria in a contaminated groundwater mediated by a genomic island in Ralstonia. *Environmental Microbiology* **5**, 163.

Mulpuru, S.K., Madhavan, M., McLeod, C.J. et al. 2017. Cardiac pacemakers: Function, troubleshooting, and management: part 1 of a 2-part series. *Journal of the American College of Cardiology* **69**, 189.

Murakami, S., Nakashima, R., Yamashita, E. et al. 2002. Crystal structure of bacterial multidrug efflux transporter AcrB. *Nature* **419**, 587.

Murray, C.J., Ikuta, K.S., Sharara, F. et al. 2022. Global burden of bacterial antimicrobial resistance in 2019: A systematic analysis. *The Lancet* **399**, 629.

Nam, H., Lewis, N.E., Lerman, J.A. et al. 2012. Network context and selection in the evolution to enzyme specificity. *Science* **337**, 1101.

Nasr, K., Viswanathan, P. and Nieder, A. 2019. Number detectors spontaneously emerge in a deep neural network designed for visual object recognition. *Science Advances* **5**, eaav7903.

Near, T.J., Dornburg, A., Kuhn, K.L. et al. 2012. Ancient climate change, antifreeze, and the evolutionary diversification of Antarctic fishes. *Proceedings of the National Academy of Sciences of the United States of America* **109**, 3434.

Nelson, G. 1993. A brief history of cardiac pacing. *Texas Heart Institute Journal* **20**, 12.

Neme, R. and Tautz, D. 2016. Fast turnover of genome transcription across evolutionary time exposes entire non-coding DNA to *de novo* gene emergence. *eLife* **5**, e09977.

Neslen, A. 10 August 2017. Monsanto sold banned chemicals for years despite known health risks, archives reveal. *The Guardian.*

Nieder, A. 2016. Representing something out of nothing: The dawning of zero. *Trends in Cognitive Sciences* **20**, 830.

Nieder, A. 2017. Number faculty is rooted in our biological heritage. *Trends in Cognitive Sciences* **21**, 403.

Nieder, A. 2018a. Evolution of cognitive and neural solutions enabling numerosity judgements: Lessons from primates and corvids. *Philosophical Transactions of the Royal Society B: Biological Sciences* **373**.

Nieder, A. 2018b. Honey bees zero in on the empty set. *Science* **360**, 1069.

Nieder, A. and Dehaene, S. 2009. Representation of Number in the Brain. *Annual Review of Neuroscience* **32**, 185.

Nishiwaki, M. 1950. On the body weight of whales. http://www.icrwhale.org/pdf/SC004184-209.pdf

Nohynek, L.J., Suhonen, E.L., Nurmiaho-Lassila, E.L. et al. 1996. Description of four pentachlorophenol-degrading bacterial strains as *Sphingomonas chlorophenolica* sp nov. *Systematic and Applied Microbiology* **18**, 527.

Norell, M.A., Clark, J.M., Chiappe, L.M. et al. 1995. A nesting dinosaur. *Nature* **378**, 774.

Notebaart, R.A., Szappanos, B., Kintses, B. et al. 2014. Network-level architecture and the evolutionary potential of underground metabolism. *Proceedings of the National Academy of Sciences of the United States of America* **111**, 11762.

Nuñez, R.E. 2017. Is there really an evolved capacity for number? *Trends in Cognitive Sciences* **21**, 409.

Nutman, A.P., Bennett, V.C., Friend, C.R.L. et al. 2016. Rapid emergence of life shown by discovery of 3,700-million-year-old microbial structures. *Nature* **537**, 535.

O'Bleness, M., Searles, V.B., Varki, A. et al. 2012. Evolution of genetic and genomic features unique to the human lineage. *Nature Reviews Genetics* **13**, 853.

O'Maille, P.E., Malone, A., Dellas, N. et al. 2008. Quantitative exploration of the catalytic landscape separating divergent plant sesquiterpene synthases. *Nature Chemical Biology* **4**, 617.

Ogburn, W.F. and Thomas, D. 1922. Are inventions inevitable? A note on social evolution. *Political Science Quarterly* **37**, 83.

Ohba, N. and Nakao, K. 2012. Sleeping beauties in ophthalmology. *Scientometrics* **93**, 253.

Ohno, S. 1970. *Evolution by Gene Duplication*. Springer, New York, NY.

Olszewski, K.L., Mather, M.W., Morrisey, J.M. et al. 2010. Branched tricarboxylic acid metabolism in *Plasmodium falciparum*. *Nature* **466**, 774.

Ossowski, S., Schneeberger, K., Lucas-Lledó, J.I. et al. 2010. The rate and molecular spectrum of spontaneous mutations in *Arabidopsis thaliana*. *Science* **327**, 92.

Ottoni, E.B. and Izar, P. 2008. Capuchin monkey tool use: Overview and implications. *Evolutionary Anthropology* **17**, 171.

Palmieri, N., Kosiol, C. and Schlotterer, C. 2014. The life cycle of *Drosophila* orphan genes. *eLife* **3**, e01311.

Panteleeva, S., Reznikova, Z. and Vygonyailova, O. 2013. Quantity judgments in the context of risk/reward decision making in striped field mice: First 'count', then hunt. *Frontiers in Psychology* **4**, 53.

Paterson, J.R., Garcia-Bellido, D.C., Lee, M.S.Y. et al. 2011. Acute vision in the giant Cambrian predator *Anomalocaris* and the origin of compound eyes. *Nature* **480**, 237.

Pecoits, E., Smith, M.L., Catling, D.C. et al. 2015. Atmospheric hydrogen peroxide and Eoarchean iron formations. *Geobiology* **13**, 1.

Phillips, M.A., Long, A.D., Greenspan, Z.S. et al. 2016. Genome-wide analysis of long-term evolutionary domestication in *Drosophila melanogaster*. *Scientific Reports* **6**, 39281.

Piantadosi, S.T. and Cantlon, J.F. 2017. True numerical cognition in the wild. *Psychological Science* **28**, 462.

Piatigorsky, J. 1998. Gene sharing in lens and cornea: Facts and implications. *Progress in Retinal and Eye Research* **17**, 145.

Piatigorsky, J. and Wistow, G.J. 1989. Enzyme crystallins: Gene sharing as an evolutionary strategy. *Cell* **57**, 197.

Pica, P., Lemer, C., Izard, W. et al. 2004. Exact and approximate arithmetic in an Amazonian indigene group. *Science* **306**, 499.

Pickrell, J. 2019. How the earliest mammals thrived alongside dinosaurs. *Nature* **574**, 468.

Pinker, S. 2007. *The Stuff of Thought: Language as a Window into Human Nature*. Penguin, London, UK.

Pinker, S. and Jackendoff, R. 2005. The faculty of language: What's special about it? *Cognition* **95**, 201.

Pinzone, P., Potts, D., Pettibone, G. et al. 2018. Do novel weapons that degrade mycorrhizal mutualisms promote species invasion? *Plant Ecology* **219**, 539.

Piperno, D.R. and Sues, H.D. 2005. Dinosaurs dined on grass. *Science* **310**, 1126.

Poolman, E.M. and Galvani, A.P. 2007. Evaluating candidate agents of selective pressure for cystic fibrosis. *Journal of the Royal Society Interface* **4**, 91.

Porter, S. 2011. The rise of predators. *Geology* **39**, 607.

Powner, M.W., Gerland, B. and Sutherland, J.D. 2009. Synthesis of activated pyrimidine ribonucleotides in prebiotically plausible conditions. *Nature* **459**, 239.

Prasad, V., Stromberg, C.A.E., Alimohammadian, H. et al. 2005. Dinosaur coprolites and the early evolution of grasses and grazers. *Science* **310**, 1177.

Pritchard, D. 1989. *The Radar War: Germany's Pioneering Achievement, 1904–45.* Patrick Stephens Limited, Sparkford, UK.

Puchner, M. 2017. *The Written World: The Power of Stories to Shape People, History, Civilization.* Random House, New York, NY.

Purnomo, A.S., Mori, T., Kamei, I. et al. 2011. Basic studies and applications on bioremediation of DDT: A review. *International Biodeterioration & Biodegradation* **65**, 921.

Quinn, D.M. 1987. Acetylcholinesterase: Enzyme structure, reaction dynamics, and virtual transition states. *Chemical Reviews* **87**, 955.

Raguso, R.A. 2008. Wake up and smell the roses: The ecology and evolution of floral scent. *Annual Review of Ecology, Evolution, and Systematics* **39**, 549.

Rajakumar, K. 2001. Infantile scurvy: A historical perspective. *Pediatrics* **108**, e76.

Ramakrishnan, V. 2019. *Gene Machine.* Basic Books, New York, NY.

Ratcliff, W.C., Denison, R.F., Borrello, M. et al. 2012. Experimental evolution of multicellularity. *Proceedings of the National Academy of Sciences of the United States of America* **109**, 1595.

Ratcliff, W.C., Fankhauser, J.D., Rogers, D.W. et al. 2015. Origins of multicellular evolvability in snowflake yeast. *Nature Communications* **6**, 1.

Ratcliff, W.C., Herron, M.D., Howell, K. et al. 2013. Experimental evolution of an alternating uni- and multicellular life cycle in *Chlamydomonas reinhardtii.* *Nature Communications* **4**, 1.

Reader, S.M. and Laland, K.N. 2003. *Animal Innovation.* Oxford University Press, Oxford, UK.

Redner, S. 2005. Citation statistics from 110 years of *Physical Review.* *Physics Today* **58**, 49.

Regnier, F.E. and Law, J.H. 1968. Insect pheromones. *Journal of Lipid Research* **9**, 541.

Reinhard, C.T., Planavsky, N.J., Olson, S.L. et al. 2016. Earth's oxygen cycle and the evolution of animal life. *Proceedings of the National Academy of Sciences of the United States of America* **113**, 8933.

Reinhardt, J.A., Wanjiru, B.M., Brant, A.T. et al. 2013. *De novo* ORFs in *Drosophila* are important to organismal fitness and evolved rapidly from previously non-coding sequences. *PLOS Genetics* **9**.

Rendic, S. and DiCarlo, F.J. 1997. Human cytochrome P450 enzymes: A status report summarizing their reactions, substrates, inducers, and inhibitors. *Drug Metabolism Reviews* **29**, 413.

Richardson, D.M. and Pyšek, P. 2006. Plant invasions: Merging the concepts of species invasiveness and community invasibility. *Progress in Physical Geography* **30**, 409.

Richardson, D.M., Williams, P.A. and Hobbs, R.J. 1994. Pine invasions in the southern hemisphere – determinants of spread and invadability. *Journal of Biogeography* **21**, 511.

Robinson, D.A., Kearns, A.M., Holmes, A. et al. 2005. Re-emergence of early pandemic *Staphylococcus aureus* as a community-acquired meticillin-resistant clone. *The Lancet* **365**, 1256.

Rodrigues, J.F.M. and Wagner, A. 2009. Evolutionary plasticity and innovations in complex metabolic reaction networks. *PLOS Computational Biology* **5**.

Rogers, J. and Gibbs, R.A. 2014. Comparative primate genomics: Emerging patterns of genome content and dynamics. *Nature Reviews Genetics* **15**, 347.

Rolls, E.T. 2000. Functions of the primate temporal lobe cortical visual areas in invariant visual object and face recognition. *Neuron* **27**, 205.

Rolls, E.T. 2012. Invariant visual object and face recognition: Neural and computational bases, and a model, VisNet. *Frontiers in Computational Neuroscience* **6**, 35.

Root-Bernstein, R.S. and Root-Bernstein, M. 1999. *Sparks of Genius: The 13 Thinking Tools of the World's Most Creative People.* Houghton Mifflin, Boston, MA.

Rose, G.J. 2018. The numerical abilities of anurans and their neural correlates: Insights from neuroethological studies of acoustic communication. *Philosophical Transactions of the Royal Society B: Biological Sciences* **373**.

Rosing, M.T. 1999. C-13-depleted carbon microparticles in >3700-Ma sea-floor sedimentary rocks from west Greenland. *Science* **283**, 674.

Ruiz-Orera, J., Hernandez-Rodriguez, J., Chiva, C. et al. 2015. Origins of *de novo* genes in human and chimpanzee. *PLOS Genetics* **11**.

Ruiz-Orera, J., Verdaguer-Grau, P., Villanueva-Canas, J.L. et al. 2018. Translation of neutrally evolving peptides provides a basis for *de novo* gene evolution. *Nature Ecology & Evolution* **2**, 890.

Sage, R.F. 2004. The evolution of C-4 photosynthesis. *New Phytologist* **161**, 341.

Sahoo, S.K., Planavsky, N.J., Jiang, G. et al. 2016. Oceanic oxygenation events in the anoxic Ediacaran ocean. *Geobiology* **14**, 457.

Samal, A., Rodrigues, J.F.M., Jost, J. et al. 2010. Genotype networks in metabolic reaction spaces. *BMC Systems Biology* **4**, 1.

Sanz, C.M. and Morgan, D.B. 2013. Ecological and social correlates of chimpanzee tool use. *Philosophical Transactions of the Royal Society B: Biological Sciences* **368**.

Sargeant, B.L., Wirsing, A.J., Heithaus, M.R. et al. 2007. Can environmental heterogeneity explain individual foraging variation in wild bottlenose dolphins (Tursiops sp.)? *Behavioral Ecology and Sociobiology* **61**, 679.

Scally, A. 2016. The mutation rate in human evolution and demographic inference. *Current Opinion in Genetics & Development* **41**, 36.

Schirrmeister, B.E., Antonelli, A. and Bagheri, H.C. 2011. The origin of multicellularity in cyanobacteria. *BMC Evolutionary Biology* **11**, 1.

Schirrmeister, B.E., de Vos, J.M., Antonelli, A. et al. 2013. Evolution of multicellularity coincided with increased diversification of cyanobacteria and the Great Oxidation Event. *Proceedings of the National Academy of Sciences of the United States of America* **110**, 1791.

Schmitt-Kopplin, P., Gabelica, Z., Gougeon, R.D. et al. 2010. High molecular diversity of extraterrestrial organic matter in Murchison meteorite revealed 40 years after its fall. *Proceedings of the National Academy of Sciences of the United States of America* **107**, 2763.

Schmitz, J.F., Ullrich, K.K. and Bornberg-Bauer, E. 2018. Incipient *de novo* genes can evolve from frozen accidents that escaped rapid transcript turnover. *Nature Ecology & Evolution* **2**, 1626.

Scholz, S.S., Reichelt, M., Mekonnen, D.W. et al. 2015. Insect herbivory-elicited GABA accumulation in plants is a wound-induced, direct, systemic, and jasmonate-independent defense response. *Frontiers in Plant Science* **6**.

Schuler, M.A. and Werck-Reichhart, D. 2003. Functional genomics of P450s. *Annual Review of Plant Biology* **54**, 629.

Schumayer, D. and Hutchinson, D.A. 2011. Colloquium: Physics of the Riemann hypothesis. *Reviews of Modern Physics* **83**, 307.

Schwab, I.R. 2012. *Evolution's Witness. How Eyes Evolved.* Oxford University Press, New York, NY.

Seehausen, O. 2006. African cichlid fish: A model system in adaptive radiation research. *Proceedings of the Royal Society B: Biological Sciences* **273**, 1987.

Segre, D., Vitkup, D. and Church, G. 2002. Analysis of optimality in natural and perturbed metabolic networks. *Proceedings of the National Academy of Sciences of the United States of America* **99**, 15112.

Sender, R., Fuchs, S. and Milo, R. 2016. Revised estimates for the number of human and bacteria cells in the body. *PLOS Biology* **14**, e1002533.

Shumaker, R.W., Walkup, K.R. and Beck, B.B. 2011. *Animal Tool Behavior. The use and manufacture of tools by animals.* The Johns Hopkins University Press, Baltimore, MD.

Siblin, E. 2010. *The Cello Suites: JS Bach, Pablo Casals, and the search for a Baroque masterpiece.* House of Anansi, Toronto, Canada.

Simonton, D.K. 1988. *Scientific Genius.* Cambridge University Press, New York, NY.

Simonton, D.K. 1994. *Greatness: Who makes history and why.* The Guilford Press, New York, NY.

Simonton, D.K. 1999a. Creativity as blind variation and selective retention: Is the creative process Darwinian? *Psychological Inquiry* **10**, 309.

Simonton, D.K. 1999b. *Origins of Genius: Darwinian perspectives on creativity.* Oxford University Press, New York, NY.

Sinatra, R., Wang, D., Deville, P. et al. 2016. Quantifying the evolution of individual scientific impact. *Science* **354**, aaf5239.

Skorupski, P., Maboudi, H., Dona, H.S.G. et al. 2018. Counting insects. *Philosophical Transactions of the Royal Society B: Biological Sciences* **373**, 20160513.

Sleep, N.H., Zahnle, K. and Neuhoff, P.S. 2001. Initiation of clement surface conditions on the earliest Earth. *Proceedings of the National Academy of Sciences of the United States of America* **98**, 3666.

Smirnova, A.A., Lazareva, O.F. and Zorina, Z.A. 2000. Use of number by crows: Investigation by matching and oddity learning. *Journal of the Experimental Analysis of Behavior* **73**, 163.

Soo, V.W.C., Hanson-Manful, P. and Patrick, W.M. 2011. Artificial gene amplification reveals an abundance of promiscuous resistance determinants in *Escherichia coli*. *Proceedings of the National Academy of Sciences of the United States of America* **108**, 1484.

Spagnoletti, N., Visalberghi, E., Verderane, M.P. et al. 2012. Stone tool use in wild bearded capuchin monkeys, *Cebus libidinosus*. Is it a strategy to overcome food scarcity? *Animal Behaviour* **83**, 1285.

Sperling, E.A., Wolock, C.J., Morgan, A.S. et al. 2015. Statistical analysis of iron geochemical data suggests limited late Proterozoic oxygenation. *Nature* **523**, 451.

Sprouffske, K., Aguílar-Rodríguez, J., Sniegowski, P. et al. 2018. High mutation rates limit evolutionary adaptation in *Escherichia coli*. *PLOS Genetics* **14**, e1007324.

Srivastava, M., Begovic, E., Chapman, J. et al. 2008. The *Trichoplax* genome and the nature of placozoans. *Nature* **454**, 955.

Stal, L.J. 2015. Nitrogen fixation in cyanobacteria. In *Encyclopedia of Life Sciences (Online)*. John Wiley & Sons, Chichester.

Stanley, S.M. 2014. Evolutionary radiation of shallow-water *Lucinidae* (Bivalvia with endosymbionts) as a result of the rise of seagrasses and mangroves. *Geology* **42**, 803.

Stepanov, V.G. and Fox, G.E. 2007. Stress-driven in vivo selection of a functional mini-gene from a randomized DNA library expressing combinatorial peptides in *Escherichia coli*. *Molecular Biology and Evolution* **24**, 1480.

Steppuhn, A., Gase, K., Krock, B. et al. 2004. Nicotine's defensive function in nature. *PLOS Biology* **2**, e217.

Stern, N. 1978. Age and achievement in mathematics: A case study in the sociology of science. *Social Studies of Science* **8**, 127.

Stoianov, I. and Zorzi, M. 2012. Emergence of a 'visual number sense' in hierarchical generative models. *Nature Neuroscience* **15**, 194.

Stokstad, E. 2007. Species conservation: Can the bald eagle still soar after it is delisted? *Science* **316**, 1689.

Strandburg-Peshkin, A., Farine, D.R., Couzin, I.D. et al. 2015. Shared decision-making drives collective movement in wild baboons. *Science* **348**, 1358.

Stromberg, C.A.E. 2005. Decoupled taxonomic radiation and ecological expansion of open-habitat grasses in the Cenozoic of North America. *Proceedings of the National Academy of Sciences of the United States of America* **102**, 11980.

Stross, R. 2007. *The Wizard of Menlo Park. How Thomas Alva Edison Invented the Modern World.* Three Rivers Press, New York, NY.

Stroud, J.T. and Losos, J.B. 2016. Ecological opportunity and adaptive radiation. *Annual Review of Ecology, Evolution, and Systematics* **47**, 507.

Studer, R.A., Rodriguez-Mias, R.A., Haas, K.M. et al. 2016. Evolution of protein phosphorylation across 18 fungal species. *Science* **354**, 229.

Supuran, C.T. and Capasso, C. 2017. An overview of the bacterial carbonic anhydrases. *Metabolites* **7**, 56.

Szostak, J.W. 2017. The origin of life on Earth and the design of alternative life forms. *Molecular Frontiers Journal* **1**, 121.

Talmy, L. 1988. Force dynamics in language and cognition. *Cognitive Science* **12**, 49.

Tattersall, I. 2009. Becoming modern *Homo sapiens*. *Evolution: Education and Outreach* **2**, 584.

Taylor, A.H., Hunt, G.R., Holzhaider, J.C. et al. 2007. Spontaneous metatool use by New Caledonian crows. *Current Biology* **17**, 1504.

Taylor, P. and Radic, Z. 1994. The cholinesterases: From genes to proteins. *Annual Review of Pharmacology and Toxicology* **34**, 281.

Tebbich, S., Taborsky, M., Fessl, B. et al. 2001. Do woodpecker finches acquire tool-use by social learning? *Proceedings of the Royal Society B: Biological Sciences* **268**, 2189.

Tebbich, S., Taborsky, M., Fessl, B. et al. 2002. The ecology of tool-use in the woodpecker finch (*Cactospiza pallida*). *Ecology Letters* **5**, 656.

Tenaillon, O., Rodriguez-Verdugo, A., Gaut, R.L. et al. 2012. The molecular diversity of adaptive convergence. *Science* **335**, 457.

Thatcher, B., Doherty, A., Orvisky, E. et al. 1998. Gustin from human parotid saliva is carbonic anhydrase VI. *Biochemical and Biophysical Research Communications* **250**, 635.

Thatje, S., Hillenbrand, C.D., Mackensen, A. et al. 2008. Life hung by a thread: Endurance of antarctic fauna in glacial periods. *Ecology* **89**, 682.

Theuri, M. 2013. Water hyacinth: Can its aggressive invasion be controlled? *Environmental Development* **7**, 139.

Thompson, D. 2017. *Hit Makers. The Science of Popularity in an Age of Distraction.* Penguin Press, New York, NY.

Thouless, C.R., Fanshawe, J.H. and Bertram, B.C.R. 1989. Egyptian vultures *Neophron percnopterus* and ostrich *Struthio camelus* eggs: The origins of stone throwing behavior. *Ibis* **131**, 9.

Tocheri, M.W., Orr, C.M., Jacofsky, M.C. et al. 2008. The evolutionary history of the hominin hand since the last common ancestor of *Pan* and *Homo*. *Journal of Anatomy* **212**, 544.

Toll-Riera, M., San Millan, A., Wagner, A. et al. 2016. The genomic basis of evolutionary innovation in *Pseudomonas aeruginosa*. *PLOS Genetics* **12**, e1006005.

Tomasello, M. 1999. The human adaptation for culture. *Annual Review of Anthropology* **28**, 509.

Tompa, P. 2012. Intrinsically disordered proteins: A 10-year recap. *Trends in Biochemical Sciences* **37**, 509.

Toth, M., Smith, C., Frase, H. et al. 2010. An antibiotic-resistance enzyme from a deep-sea bacterium. *Journal of the American Chemical Society* **132**, 816.

Tripathi, S., Kloss, P.S. and Mankin, A.S. 1998. Ketolide resistance conferred by short peptides. *Journal of Biological Chemistry* **273**, 20073.

True, J.R. and Carroll, S.B. 2002. Gene co-option in physiological and morphological evolution. *Annual Review of Cell and Developmental Biology* **18**, 53.

Tully, J. 2011. *The Devil's Milk. A Social History of Rubber.* Monthly Review Press, New York, NY.

Turnbull, S. 2003. *Genghis Khan and the Mongol Conquests 1190–1400.* Osprey Publishing, Oxford, UK.

Tyndall, J. 1863. XXXI. Remarks on an article entitled 'Energy' in 'Good words'. *The London, Edinburgh, and Dublin Philosophical Magazine and Journal of Science* **25**, 220.

Uauy, R., Gattas, V. and Yanez, E. 1995. Sweet lupins in human nutrition. In *Plants in Human Nutrition* (ed. A.P. Simopoulos), p. 75. Karger Publishers, Basel, Switzerland.

Ukuwela, K.D.B., de Silva, A., Mumpuni et al. 2013. Molecular evidence that the deadliest sea snake *Enhydrina schistosa* (Elapidae: Hydrophiinae) consists of two convergent species. *Molecular Phylogenetics and Evolution* **66**, 262.

Urayama, A. 1971. Unilateral acute uveitis with retinal periarteritis and detachment. *Rinsho Ganka (Japanese Journal of Clinical Ophthalmolology)* **25**, 607.

Van Calster, B. 2012. It takes time: A remarkable example of delayed recognition. *Journal of the American Society for Information Science and Technology* **63**, 2341.

van der Heide, T., Govers, L.L., de Fouw, J. et al. 2012. A three-stage symbiosis forms the foundation of seagrass ecosystems. *Science* **336**, 1432.

Van Raan, A.F. 2004. Sleeping beauties in science. *Scientometrics* **59**, 467.

van Schaik, C.P., Ancrenaz, M., Borgen, G. et al. 2003a. Orangutan cultures and the evolution of material culture. *Science* **299**, 102.

van Schaik, C.P., Fox, E.A. and Fechtman, L.T. 2003b. Individual variation in the rate of use of tree-hole tools among wild orang-utans: Implications for hominin evolution. *Journal of Human Evolution* **44**, 11.

Verguts, T. and Fias, W. 2004. Representation of number in animals and humans: A neural model. *Journal of Cognitive Neuroscience* **16**, 1493.

Vogel, A.C., Petersen, S.E. and Schlaggar, B.L. 2014. The VWFA: It's not just for words anymore. *Frontiers in Human Neuroscience* **8**, 88.

Voje, K.L., Starrfelt, J. and Liow, L.H. 2018. Model adequacy and microevolutionary explanations for stasis in the fossil record. *The American Naturalist* **191**, 509.

Wagner, A. 2012. The role of randomness in Darwinian Evolution. *Philosophy of Science* **79**, 95.

Wagner, A. 2019. *Life Finds a Way: What Evolution Teaches Us About Creativity.* Basic Books, New York, NY.

Wagner, A. and Rosen, W. 2014. Spaces of the possible: Universal Darwinism and the wall between technological and biological innovation. *Journal of the Royal Society Interface* **11**, 20131190.

Wagner, A. 2022. Competition for nutrients increases invasion resistance during assembly of microbial communities. *Molecular Ecology* **31**, 4188.

Waller, J. 2004. *Leaps in the Dark: The making of scientific reputations.* Oxford University Press, Oxford.

Wang, J.J., Odic, D., Halberda, J. et al. 2016. Changing the precision of preschoolers' approximate number system representations changes their symbolic math performance. *Journal of Experimental Child Psychology* **147**, 82.

Wang, M., Jiang, Y.-Y., Kim, K.M. et al. 2011. A universal molecular clock of protein folds and its power in tracing the early history of aerobic metabolism and planet oxygenation. *Molecular Biology and Evolution* **28**, 567.

Ward, A. 1986. *John Keats. The Making of a Poet.* Farrar, Straus and Giroux, New York, NY.

Ward, C., Henderson, S. and Metcalfe, N.H. 2013. A short history on pacemakers. *International Journal of Cardiology* **169**, 244.

Watanabe, K., Urasopon, N. and Malaivijitnond, S. 2007. Long-tailed macaques use human hair as dental floss. *American Journal of Primatology* **69**, 940.

Weaver, L.H., Grutter, M.G., Remington, S.J. et al. 1985. Comparison of goose-type, chicken-type, and phage-type lysozymes illustrates the changes that occur in both amino acid sequence and 3-dimensional structure during evolution. *Journal of Molecular Evolution* **21**, 97.

Weikert, C. and Wagner, A. 2012. Phenotypic constraints and phenotypic hitchhiking in a promiscuous enzyme. *The Open Evolution Journal* **6**, 14.

Weng, J.-K. and Noel, J. 2012. The remarkable pliability and promiscuity of specialized metabolism. In *Cold Spring Harbor Symposia on Quantitative Biology*, p. 309. Cold Spring Harbor Laboratory Press.

Weng, J.K., Philippe, R.N. and Noel, J.P. 2012. The rise of chemodiversity in plants. *Science* **336**, 1667.

Werneburg, I., Laurin, M., Koyabu, D. et al. 2016. Evolution of organogenesis and the origin of altriciality in mammals. *Evolution & Development* **18**, 229.

Westfall, C.S., Zubieta, C., Herrmann, J. et al. 2012. Structural basis for prereceptor modulation of plant hormones by GH3 Proteins. *Science* **336**, 1708.

Whiten, A., Goodall, J., McGrew, W.C. et al. 1999. Cultures in chimpanzees. *Nature* **399**, 682.

Whiten, A., Goodall, J., McGrew, W.C. et al. 2001. Charting cultural variation in chimpanzees. *Behaviour* **138**, 1481.

Whiten, A., Horner, V. and de Waal, F.B.M. 2005. Conformity to cultural norms of tool use in chimpanzees. *Nature* **437**, 737.

Whittaker, R.J. and Fernandez-Palacios, J.M. 2007. *Island Biogeography: Ecology, Evolution, and Conservation.* Oxford University Press, Oxford, UK.

Whittington, H.B. 1975. Enigmatic animal *Opabinia regalis,* middle Cambrian, Burgess shale, Britisch Columbia. *Philosophical Transactions of the Royal Society of London Series B-Biological Sciences* **271**, 1.

Wiberg, R.A.W., Halligan, D.L., Ness, R.W. et al. 2015. Assessing recent selection and functionality at long noncoding RNA loci in the mouse genome. *Genome Biology and Evolution* **7**, 2432.

Wilson, B.A., Foy, S.G., Neme, R. et al. 2017. Young genes are highly disordered as predicted by the preadaptation hypothesis of *de novo* gene birth. *Nature Ecology & Evolution* **1**.

Wimpenny, J.H., Weir, A.A.S., Clayton, L. et al. 2009. Cognitive processes associated with sequential tool use in New Caledonian crows. *PLOS One* **4**.

Wistow, G.J. and Piatigorsky, J. 1988. Lens crystallins – the evolution and expression of proteins for a highly specialized tissue. *Annual Review of Biochemistry* **57**, 479.

Wood, R. and Erwin, D.H. 2018. Innovation not recovery: Dynamic redox promotes metazoan radiations. *Biological Reviews* **93**, 863.

Wood, R., Liu, A.G., Bowyer, F. et al. 2019. Integrated records of environmental change and evolution challenge the Cambrian explosion. *Nature Ecology & Evolution* **3**, 528.

Wood, R.A., Poulton, S.W., Prave, A.R. et al. 2015. Dynamic redox conditions control late Ediacaran metazoan ecosystems in the Nama Group, Namibia. *Precambrian Research* **261**, 252.

Wynn, K. 1992. Addition and subtraction by human infants. *Nature* **358**, 749.

Xiao, W., Liu, H., Li, Y. et al. 2009. A rice gene of *de novo* origin negatively regulates pathogen-induced defense response. *PLOS One* **4**, e4603.

Xu, J., Chmela, V., Green, N.J. et al. 2020. Selective prebiotic formation of RNA pyrimidine and DNA purine nucleosides. *Nature* **582**, 60.

Xu, X. and Norell, M.A. 2004. A new troodontid dinosaur from China with avian-like sleeping posture. *Nature* **431**, 838.

Xu, X., Zhou, Z.H., Dudley, R. et al. 2014. An integrative approach to understanding bird origins. *Science* **346**, 1341.

Yamaguchi, N., Ichijo, T., Sakotani, A. et al. 2012. Global dispersion of bacterial cells on Asian dust. *Scientific Reports* **2**.

Yang, H., Jaime, M., Polihronakis, M. et al. 2018. Re-annotation of eight *Drosophila* genomes. *Life Science Alliance* **1**, e201800156.

Yeung, A.W.K. and Ho, Y.S. 2018. Identification and analysis of classic articles and sleeping beauties in neurosciences. *Current Science* **114**, 2039.

Yona, A.H., Alm, E.J. and Gore, J. 2018. Random sequences rapidly evolve into *de novo* promoters. *Nature Communications* **9**, 1530.

Yu, J.-F., Cao, Z., Yang, Y. et al. 2016. Natural protein sequences are more intrinsically disordered than random sequences. *Cellular and Molecular Life Sciences* **73**, 2949.

Zachar, I. and Szathmary, E. 2017. Breath-giving cooperation: Critical review of origin of mitochondria hypotheses. Major unanswered questions point to the importance of early ecology. *Biology Direct* **12**, 1.

Zhang, S.C., Wang, X.M., Wang, H.J. et al. 2016. Sufficient oxygen for animal respiration 1,400 million years ago. *Proceedings of the National Academy of Sciences of the United States of America* **113**, 1731.

Zhao, L., Saelao, P., Jones, C.D. et al. 2014. Origin and spread of *de novo* genes in *Drosophila melanogaster* populations. *Science* **343**, 769.

INDEX

References to images are in *italics*; references to notes are indicated by n.

absorption refrigeration 206–7
adaptive radiation 18–21, 33, 243 n. 10
Aesop 161
Africa, *see* East Africa
agriculture 10, 39, 90, 199–200
and ants 6, 58
alexia 177–8
algae 68–9
alkaloids 116
amber 17
amino acids 31, 40–1, 100–2, 103
ammonia 90
analogies 10, 185–7, 192, 196
ancient Egypt 173, 197, 273 n. 1
Andes Mountains 6, 22–3
animals 39, 45, 60–7, 168–9
and numerosity 169–71
and tools 149–51, 152–64

Anomalocaris 63, 66
Anson, George 202
ant lions 158
anteaters 52, 53
antibiotics 5, 6, 7, 36–7, 101–2
and resistance 96–100, 104–5, 106–10
and soils 112–13
antiscorbutics 204–6
ants 2, 6, 7, 57–8, 149, 227
aphids 58
approximate number system 167
arabinose 87
Archaeopteryx 54–5
Arendt, Hannah 232
Aroclor 77
art 229–30
arthropods 62
artificial intelligence (AI) 171, 220

asteroids 50, 56, 228
atoms 10, 185
ATP synthase 42, 44

baboons 169
Bach, Johann Sebastian 11, 231
Backus, John 201
bacteria 5, 7–8, 59, 111, 225
    and antibiotics 95–100,
        104–5, 106–10, 112–13
    and biofuels 87
    and gene transfer 75–7
    see also cyanobacteria; Esche-
        richia coli; Sphingomonas
        chlorophenolicum
Bakken, Earl E. 210
beavers 52, 53
bees 170, 171
beneficial mutation 32–3, 36
beta-lactamases 104
Bichat, Marie-François-Xavier
    209
big data 219–20
bile acids 105
biofuels 87
biosynthetic enzymes 116
biotechnology 8
birds 7, 53–7, 227–8
    and tools 157–60, 161–2
    see also crows
bitter vine (Mikania micrantha)
    119
black bears 170
blue whales 4, 50
body language 168–9
brain, the 10, 192–6
    and images 179–83

and mathematics 173–6
and reading 178–9
Broglie, Louis de 185
burrowing animals 66

C_4 photosynthesis 45–7, 48, 49
cacti 49, 228
caffeine 116, 222
calcium signalling 216
Cambrian explosion 62–4, 65–7,
    250 n. 51
camouflage 25, 150
Campbell, Donald 237
captivity 159–62
capuchin monkeys 151, 153–4,
    160, 162–3
car industry 235
carbon dioxide 42–3, 46–7, 48,
    49, 72, 103
cardiac glycosides 18
cardiac pacemakers 208–12, 227
Carnot, Sadi 186
Carroll, Lewis: Through the Look-
    ing-Glass 99–100
Castorocauda 52
caterpillars 17–18
causation 190
Cavendish, Henry 221
chameleons 29, 223
Chandrasekhar, Subramanyan
    237
Changizi, Mark 180, 181–3
chariots 197, 198
chemical weapons 17–19,
    118–19
chemofossils 72–3
chimpanzees 3, 130, 131

and tools 149, 152–3, 155–6, 163
China 51, 198
Chlorella vulgaris (algae) 68–9
chlorobenzene 77
chromosomes 30–1
cichlids 24–5, 28, 72–3, 243 n. 10
clams 2, 59–60
Clever Hans 168–9
climate change 47–9, 56, 58–9, 228
co-option 224–5
coloration 24–5
computers 82–6, 219–20
conceptual metaphors 188
consortia 88–9, 93–4
convergent evolution 53
Cook, Capt James 275 n. 15
cooling systems 224
crops 6, 39, 90
cross-modality 169–70
crows 3, 9, 170, 171
and tools 157, 159–60, 161–2
crystallins 114–15, 121
Csikszentmihalyi, Mihaly 237–8
culture, *see* human culture
cyanobacteria 7, 78–9, 89, 90
cyanogenic molecules 115
cystic fibrosis 229
cytochromes P450: 117, 128

Da Gama, Vasco 202
Daphnia (water flea) 143–5
Darwin, Charles 123, 186, 216
Origin of Species by Means of Natural Selection 83, 119–20, 215

Dawkins, Richard 127, 260 n. 6
DDT 77
de novo genes 124–5, 128–9, 133–9
De Vries, Hugo 216
deep learning 220
deep neural networks 171–2
defence chemicals 115–16
Dehaene, Stanislas 174, 183
Deinopsis (gladiator spider) 149, *150*
Déjerine, Joseph-Jules 177
deserts 28–9, 48
detritus-feeders 19
Diania cactiformis (walking cactus) 62
Dickinson, Emily 11, 231
Dickson, Leonard 219
dinosaurs 2, 39–40, 53–4
and extinction 50–1, 52, 56, 227–8
disease, *see* heart disease; malaria; scurvy
DNA 4–5, 8–9
and bacteria 74–6, 98
and cells 68
and duplications 125–9
and genes 42, 43, 113–14, 132, 138–9
and mutation 40–1, 109, 124, 142–3
and non-coding (junk) 133–6
and ribosome 129–30
and sequencers 82
and strings 30–3
and transcription 139–41

dolphins 3, 9, 156–7
Drosophila melanogaster (fruit
fly) 33, 131, 135–6
drought 2, 45

E. coli, see Escherichia coli
(E. coli)
East Africa 23–4, 28
Eastgate Centre (Zimbabwe)
224
ecdysone 116
echinoderms 62
Edison, Thomas 201, 233–4, 278
n. 23–4
efflux pumps 104–6, 257 n. 22–3
Einstein, Albert 185, 275 n. 17
El Greco 11, 231
electricity 233–4, 278 n. 23–4
electrons 185
Elmqvist, Rune 208–9, 211
energy 7, 12
Enhydrina schistosa (sea snake)
222, 223
environment 6, 143–6, 225–6
enzymes 31, 61, 72, 81–4, 89–90,
101–4, 106
and protein 41–2, 75, 76, 80
see also promiscuous en-
zymes; RuBisCO
Eomaia scansoria 51
Escherichia coli (E. coli) 33, 34,
35–6, 76, 79–80, 81
and antibiotic resistance 106,
108–9
and enzymes 110–11
and genes 134, 135
and metabolism 84, 85–6

Etruscan shrew 50
evolution 3–6, 222–4
and animals 60–7
and antibiotics 99–100
and birds 53–7
and chemistry 77–81
and clams 59–60
and exaptation 119–21
and experimental 30–7
and fish 24–5, 26–8
and genomes 8–9, 130–3
and gradualism 123–5
and grasses 1–2, 44–5, 48–9
and insects 57–8
and lupins 23
and mammals 50–3
and molecules 7–8
and multicellular organisms
89
and technology 199–201
and teeth 20–1
and toxic excretions 18–19
exaptation 119–21
extinction 7, 224
and dinosaurs 50–1, 52, 53,
56, 227–8
eyes 114–15, 222, 216, 217

Farris, Floyd 213
fatty acids 71
fish 24–8, 119–20
and Antarctic 58–9
and benthic 249 n. 45
and numerosity 170
fitness 83–4
Fleming, Alexander 98–9, 217
flies 33, 131, 133–4, 135–6

flying squirrels 52, 53
fMRI (functional magnetic resonance imaging) 174
Ford, Henry 235
forests 39, 48, 56, 58, 154, 162
fossils 5, 29, 60
  and amber 17
  and ants 57
  and birds 54–5, 56
  and dinosaurs 39–40
  and grasses 2
  and mammals 51
  and multicellular organisms
    68
  and sticklebacks 26–8
  see also chemofossils; trace
    fossils
fracking 212–13, 227
free radicals 31
frogs 170
fructose 79, 86
fruit flies 33, 131, 133–4, 135–6
Fruitafossor 52
fungi 6, 58, 98–9

Gauss, Carl Friedrich 222
General Electric (GE) 207
generative metaphors 189–90
genes 5, 8–9, 130–3
  and chromosomes 30–1
  and Daphnia 143–5
  and de novo 124–5, 128–30,
    133–9
  and DNA 125–8
  and environment 145–6
  and horizontal transfer 74–7,
    96

  and Kyoto Encyclopaedia 84
  and Mendel 215–16
  and molecular clocks 41–4
  and mutations 140–1
  and orphans 132–3
  and protein 105–6
  and 'selfish' 260 n. 6
  and transcriptional regulators
    113–14
gliding mammals 52
glucose 35, 79, 86
Goodyear, Charles 200
Gould, Stephen Jay 120–1
  Wonderful Life 30, 62
gradualism 123–5
grasses 1–2, 7, 38–40, 44–7,
    48–9, 228; see also seagrass
Grassmann, Hermann 226–7
  Die Ausdehnungslehre 11,
    12
grazing animals 39, 45
green flies (Chloropidae) 20
grid cells 192–4
group theory 219
Gutenberg, Johannes 224

Haber-Bosch process 90
Hamilton, Bill 68
hands 184
Hangul writing 227
Hardy, G. H. 218–19
Harvey, William 186
heart disease 208–12
Henkelotherium 51
herpes 216
Hertz, Heinrich 219
horizontal drilling 213, 227

horizontal gene transfer 74–7, 96
housekeeping genes 42
Hoyle, Fred 123
Hülsmeyer, Christian 227
human culture 3–4, 6, 9–12
   and grasslands 38–9
   and mathematics 172–5
   and numerosity 165–8, 171,
     172
   and skills 183–4
   and tools 151
Hume, David 190
Hunt, Gene 28
Hyman, Albert 209, 212
hypocone 20–1

indigenous people, see Mundurukú people; Olmec people;
   Yanomami Indians
industrial revolution 200
infants 165–6
innovation 2–6, 9–12
   and agriculture 199–200
   and birds 55–6, 57
   and chemistry 78–9
   and co-option 224–6
   and genomes 8–9
   and grasses 44–7, 48–9
   and molecules 7–8
   and multicellular organisms
     67
   and nervous systems 151–2
   and social learning 154–9
   and technology 200–2
   see also key innovations
insects 19–20; see also ants; caterpillars; flies; spiders

intraparietal sulcus 174–5
invariance 178–9
invasive species 8, 117–19

Jacob, François 124–5, 128–9,
   134, 137, 139
Johnson, Lyndon B. 234
Juramaia 51
Jurassic period 51, 54

Karve, Shraddha 107–9
Keats, John 11, 230–1, 232, 238
key innovations 18–21, 241 n. 3
knowledge tradition 153–4
Köhler, Wolfgang 149
kudzu 117

lactose 121
Lakoff, George 10, 188
Lancaster, Capt James 274 n. 14
language 176–7, 180–3, 188–92
Larsson, Arne 208–9, 211, 212
latex 17–19, 21, 115, 200
Lechuguilla Cave (New Mexico)
   98
lenses 114–15
Lenski, Richard 33, 35
Libby, Eric 91–3
lighting 201, 233–4, 278
   n. 23–4
lignin 2, 45
Lillehei, C. Walton 210
Lind, James 203, 204–5
line junctions 180
linear algebra 11–12, 226–7
lions 170
literature 230–1, 232

Lucinidae (salt water clams) 2, 59–60
lung fish 119–20
lupins 22, 23, 25, 72
lysozyme 121

macaques 151, 157, 161, 169–70
Macaulay, Thomas 221
macroevolutionary lag 57
Magellan, Ferdinand 202
Mahfouz, Naguib 237
malaria 228–9
Malawi, Lake 6, 23–5, 28
mammals 2, 7, 20–1, 50–3, 227
Marche, Stephen 232
Marconi, Guglielmo 219
marketing 207–8
mathematics 3, 4, 10, 165–6, 167–8
   and the brain 174–6
   and discoveries 217–19, 221–2
   and linear algebra 11–12, 226–7
   and origins 172–4
   see also numerosity
Maxwell, James Clerk 219
Mayer, Julius Robert 12
Mei long 55
Melville, Herman 11, 231
Mendel, Gregor 3, 4, 215–16
Mentral, George de 224
meristems 45, 47
Merton, Robert K. 201–2, 223
metabolic pathways 76–88
metabolism 7, 92–4; *see also* metabolic pathway

metaphors 10, 185–6, 188–92, 196
metatools 160
meteorites 70, 71, 72
mice 140, 170
microbes 88–9
Microraptor 56, 249 n. 36
milkweeds 17–18
Miller, Stanley 70
Milner, Brenda 237
Miocene period 48
mitochondria 252 n. 15
Mitter, Charles 20
molecular clocks 40–4
molecules 7–8, 71, 82, 225
   and nutrients 79–80
   and plants 115–16
   see also DNA
molluscs 62
monarch butterfly caterpillars 17–18
Mongols 38
monkeys 9, 175; *see also* capuchin monkeys; macaques
Morris, Simon Conway 223
mountain ranges 48; *see also* Andes Mountains
moveable type 224
multicellular organisms 67–70, 89–92
multiple discoveries 200–2, 221–4
Mundurukú people 173
music 185, 231
mutation 31–3, 35–7, 130–1, 246 n. 6
   and disease 228–9

and DNA 40-1, 109, 124, 125-8, 142-3
and genes 140-1
natural selection 130-1
navigation 192-5
nervous systems 3, 102-3, 151-2
neuronal recycling 174
neurons 171-2, 175-6, 192-5
and reading 178-9, 183
neutral mutation 41
nicotine 116
nitrogen 22, 90
nomadic life 38
notothenioids (Antarctic fish) 58, 249 n. 45
numerosity 166-8, 169-72
nutrients 79-80, 86, 110-11

oblivion 229-32
Ochromonas vallescia 68-9
octopuses 150
Offner, Frank 237
Ohm's law 221
Olmec people 197-8, 234
Opabinia 62-3
open reading frames 135-6, 138
orangutans 152, 154, 163
origin of life 70-3, 139
orphan genes 132-3
Oviraptor 55
oxcarts 197, 198
oxygen 46, 64-5, 66, 78-9

palaeontology 25-8, 51
páramo habitat 22-3
pendulum clocks 200
penicillin 96, 97, 98-9, 104, 227

and World War II 217, 218
pentachlorophenol 74, 80, 89
peptides 136-7
phosphate 112
phosphoinositides 216, 217
phosphoswitches 142-3
photosynthesis 7, 45-7, 78-9
physics 219
phytolites 39-40
phytophagy 19-21
pine trees 118-19
Pinker, Steven 188
place cells 192
plant-feeders 19-21; see also grazing animals
platypuses 29, 223, 277 n. 7
Pliny the Elder 161
point mutation 32, 40-1
politics 227
pollen 40
Pre-Columbian Amerindians 200
Precambrian period 65
predation 66
primates 131; see also chimpanzees; monkeys
printing press 224
promiscuous enzymes 103-4, 106, 110-13, 114, 116-17, 121-2, 225
proteins 7-8
and antibiotics 36-7, 100-1, 104
and efflux pumps 105-6
and exaptation 121
and folds 31-2
and mutation 40-1
and peptides 136-7

and phosphate 142
and transcriptional regulators
113–14, 134–5
Proterozoic period 60, 61
prototypes 188–9, 190
punctuated equilibria 244 n. 20

quantum theory 185

radar 201, 224, 227
radiation 10
radio 219
railroads 234, 235
Ratcliffe, Will 69
rats 140, 192–3, 194–5
reading 176–9, 183
recombination 76–7
refrigeration 206–7
replication 34–5
reproduction 71
reptiles 55
resins 17–19, 115
respiration 79
retinal disease 216, 217
ribosome 129–30
rice plants 117, 128, 136, 143
RNA 31, 40–1, 70, 113, 129,
137
road networks 235
rooks 161–2
rubber 200, 274 n. 5
RuBisCO 42–3, 44, 46–7
Russell, Bertrand 172

sailors 202–5
salt water clams 2, 59–60
Schrödinger, Erwin 185

scientific publications 213–17,
236
screw press 224
scurvy 12, 202–6, 274 n.
14–15
sea otters 150–1, 152, 162
sea urchins 150
seagrass 59–60
Senning, Åke 208–9, 211
shapes 180–3
shelled animals 66
shrews 50, 53
sickle cell anaemia 229
silicon dioxide 2, 45
skeletons 51
snakes 222, *223*
social learning 152–9, 208
sound 186–7, 194–6
source domains 187, 188
space 190–6
Sphingomonas chlorophenolicum
74–5, 76, 77, 79
spiders 149, *150*, 170
spotted knapweed 119
Stern, Richard 237
sticklebacks 25–8
Stokes-Adams attacks 208–9
streptomycin 6
stroke patients 176, 177–8
subitising 167, 169
superbugs 8
swim bladders 119–20
Szilard, Leo 275 n. 17

Tanganyika, Lake 23
tannins 116
target domains 187, 188

technology 9–10, 12, 199–201, 207–8
teeth 20–1, 45, 52
telegraph 200–1, 219
thermometers 200
Thompson, Derek: *Hitmakers* 233
Thoré-Bürger, Théophile 230
Thoreau, Henry David 11
  Walden 230, 232
time 190–2
Toll-Riera, Macarena 111
tools 9, 149–51, 152–64, 265 n. 2
toxic excretions 17–19
toxins 7, 8, 89
trace fossils 65–6
transport 197–8, 234–5
tree-living mammals 51, 52
Trichoplax adhaerens 61
tuberculosis 229

vacuum cleaners 207–8
Van Gogh, Vincent 11, 231
Velcro 224
Vermeer, Johannes 3, 4, 11, 229–30, 232
vertebrates 62
Victoria, Lake 23, 118
vision 178–80; *see also* eyes
Vitruvius 186

Volaticotherium antiquus 52
Vrba, Elisabeth 120–1
vulcanisation 200
vultures 158–9

washing machines 208
wasps 149–50
water 2, 49, 186–7
water fleas 143–5
water hyacinths 8, 117–18
water-living mammals 52
Weber's law 167, 169, 172
Welwitschia mirabilis 28–9, 30, 223
wheel 3, 12, 197–9, 234
wheelbarrows 197, 198, 224
Whiten, Andrew 152–3
Whittington, Harry 62
Wiwaxia 62
woodpecker finches 159, 162
word reading area 178–9
writing 3, 10, 180–3, 227

xylose 87

Yanomami Indians 7, 95, 96–7, 121
yeast 69, 132, 135, 142–3

zeitgeist 227